GEOLOGY
along
TRAIL RIDGE ROAD

Rocky Mountain National Park
Colorado

by

Omer B. Raup

Photography by Omer B. Raup and others
Pen and ink drawings by Omer B. Raup and Arthur L. Isom

Prepared by the U.S. Geological Survey
in cooperation with the National Park Service
Published by Falcon Press and the Rocky Mountain Nature Association
John Gunn, publications coordinator

© 1996 by the Rocky Mountain Nature Association
Published with Falcon Press Publishing Co., Inc., Helena and Billings, Montana.
Printed in Hong Kong.
10   9   8   7   6   5   4   3   2   1

Library of Congress Catalog Card Number: 96-84363
ISBN 1-56044-481-9

Cover photograph:
Gorge Lakes are nestled in the high peaks across from Rock Cut.
Photo by James Frank.

Back cover photograph:
These pegmatite spires are a familiar landmark at Milner Pass.
Photo by Omer Raup.

Contents

THE TOUR

Foreword

Imagine yourself a prospector. In the late 1850s, or perhaps the 1870s, you traveled across the Great Plains. Following the gold rush, you and your companions suffered the pangs of thirst and hunger. Perhaps you encountered some Indians. But your vision remained steady. You had dreams filled with wealth: gold and silver.

Hidden in these mountains, you were convinced, awaited a treasure. With a pick and a shovel, you intended to find it. You had heard of other Colorado places where people had struck it rich. The names were magic, places called Central City, Black Hawk, Cripple Creek, Leadville. Gold, more gold, silver and gold.

Someone should have put up a sign near the foothills as a warning to prospectors: Be prepared to fail.

For anyone who tramped across Trail Ridge in the land which would become Rocky Mountain National Park, looking for gold proved luckless. Prospectors left without riches, and they left barely a trace of their comings and goings. But that was good news for conservationists. Rocky Mountain National Park was created in 1915.

Rocky Mountain National Park owes its origins to its geology. It boasts magnificent mountain scenery. Its high mountain lakes and streams gained the instant approval of fishermen. Vacationers by the thousands came to explore its high country trails and enjoy the alpine vistas. The glacier-carved valleys were sights to behold.

Prospectors of a different ilk began tramping through the mountains. They were people interested in learning whatever they could about nature. It became the task of the National Park Service, especially its band of ranger-naturalists, to explain as much as they could about the flora, fauna, and natural history of this region.

One of the first questions visitors were likely to ask centered on the geologic story of the park. In fact, Dorr Yeager, the first chief park naturalist, considered the subject so important that the first publication produced by the Rocky Mountain Nature Association (founded in 1931) was a booklet entitled *Geology of the Park*. Since that time, hundreds of naturalist programs have addressed subjects such as glaciation, geomorphology, the brief mining history of Lulu City, and the making of Moraine Park by glaciers.

The objective of any naturalist is to educate park visitors, to open their eyes to the natural wonders of Rocky Mountain National Park. To see the park through a new set of eyes is special, indeed. What a treat it would be to see the park through the eyes of a botanist, a zoologist, or any specialist who has studied the landscape so thoroughly that the mysteries of the Earth are completely revealed. Such guides are rare.

In explaining geology as it appears along Trail Ridge Road, we are fortunate to have found a guide and naturalist in Dr. Omer B. Raup. Here, he offers a look at the park through the eyes of a highly-trained and well-experienced geologist.

A graduate of American University (B.S.) and the University of Colorado (Ph.D), Dr. Raup also was a Fulbright Scholar at the University of Edinburgh, Scotland. He has been a research geologist with the U.S. Geological Survey for forty-one years, and a frequent visitor to Rocky Mountain National Park for more than thirty-five years. His *Geology Along Going-to-the-Sun Road, Glacier National Park* (1983) has introduced thousands of visitors to the geological wonders of the northern Rockies.

Like the prospectors of yesterday, we still dream of wealth. But today, on our trip across the Rockies, the nuggets we gain from Dr. Raup are gems of knowledge and geology. They should prove to be every bit as good as gold. ∎

C. W. Buchholtz
EXECUTIVE DIRECTOR
ROCKY MOUNTAIN NATURE ASSOCIATION

Preface

Geologic studies in the area that is now Rocky Mountain National Park got underway in the late 1800s and early 1900s as the national quest for mineral deposits with economic value, and the search for knowledge about the processes of uplift that formed the Rocky Mountains, began in earnest. In the years since that time, many studies of the region's Proterozoic igneous and metamorphic rocks, Tertiary granites and volcanic rocks, structural geology, glacial geology, and geomorphology have been carried on by college and university professors, their students, and members of the U.S. Geological Survey. A geologic map of Rocky Mountain National Park published in 1990 by William A. Braddock and James C. Cole summarizes much of what is known today about the park's geology.

The work of these researchers and many other geologists have proved invaluable during the writing of this tour guide. My thanks go to William A. Braddock and William C. Bradley, former professors of mine at the University of Colorado, and to Richard F. Madole, Douglass E. Owen, and Gerald M. Richmond, my colleagues at the U.S. Geological Survey, for generously sharing their knowledge of Rocky Mountain National Park geology. Thanks also to William A. Braddock, William C. Bradley, Alfred L. Bush, Donald L. Gautier, Douglass E. Owen, and Carl L. Rich for road-testing the book. Additional thanks are due the illustrators. The generalized drawing of the panoramic view from Forest Canyon Overlook (before glaciation) was done by Carol A. Quesenberry; drafting and special graphics were done by Dennis L. Welp; and Arthur L. Isom did the wildlife sketches.

Last, I offer special thanks to my wife, Phyllis H. Raup, who encouraged me through this project and reviewed and edited the manuscript at various stages of preparation. ■

Omer B. Raup
FEBRUARY 1996

1

Introduction

This guide to the geology along spectacular Trail Ridge Road describes many of the land features of Rocky Mountain National Park. The book is designed for park visitors who have little or no background in geology. It provides information for those who want to know more about the park's rocks, and helps visitors appreciate the natural history of this scenic and geologically outstanding area. For those unfamiliar with geologic terms, it also offers definitions and additional information in the glossary on pages 67 through 70.

GEOLOGY

Geology is the study of the Earth. Although this includes rocks, mountains, and natural forces (such as earthquakes and volcanic eruptions), geology also is the study of minerals, fossils, and the processes of weathering and erosion. Its time span is nearly too long to imagine, beginning when Earth was formed about 4.5 billion years ago and continuing today.

This self-guided geologic tour of Trail Ridge Road showcases the results of several major rock- and mineral-forming processes that took place in what we now call Rocky Mountain National Park hundreds of millions of years ago. Evident along this high scenic road are ancient mountains that formed more than a billion years ago, as well as rocks and mountains formed by later volcanic eruptions. More recently in geologic time, although still thousands of years ago, these mountains were sculpted by massive Ice Age glaciers. The landforms continue to be eroded today by water, wind, and ice.

Clues to the geologic events of the distant past are found in the rocks themselves. The three major types of rocks that make up Earth's crust are igneous, sedimentary, and metamorphic rocks. These various types result from different geologic processes.

Igneous rocks form when minerals crystallize from molten (liquid) rock, called magma. For the most part, igneous rocks are either intrusive or extrusive. Intrusive igneous rocks result when magmas solidify below the surface of the Earth, some at great depths. As the magma slowly cools, it crystallizes into rocks such as granite. Examples of intrusive igneous rocks in Rocky Mountain National Park are the Silver Plume granite found at Rainbow Curve (page 29) and the Tertiary Age granites of the Never Summer Mountains (page 58).

Extrusive igneous rocks occur when molten rock erupts onto the Earth's surface. These eruptions can take the form of liquid lava flows, such as those from Hawaiian volcanoes, or they may appear as volcanic ash, such as that which exploded from Mount Saint Helens in Washington State. An example of a once-liquid lava flow, now solidified, can be found near the Bowen-Baker Trailhead (page 64). Explosive ash-flow deposits are detailed at Lava Cliffs (page 44).

Sedimentary rocks form when sediments or chemicals deposited in seas, lakes, or streams harden into rock. Common examples of these rocks are sandstones, shales, and limestones.

Metamorphic rocks are products of heat and pressure that come from forces within the Earth's crust. The extreme temperatures and stresses within the crust produce new forms of rock from what had previously been igneous or sedimentary rocks. Examples of metamorphic rocks are slate (the metamorphosed form of shale) and marble (the metamorphosed form of limestone).

Most of the rocks in Rocky Mountain National Park are igneous or metamorphic, formed during mountain-forming processes that shaped the area.

GEOLOGIC TIME

To describe events over the long span of the Earth's natural history, geologists have devised their own time scale. The concept of geologic time helps these researchers keep track of the sequence of events. The time scale is used like a calendar. Major divisions of time, called eons, show up in the left-hand column. Smaller divisions of time, called eras, periods, and epochs, take up the next three columns. The millions of years leading to the present appear in the center of the chart (see next page).

During most of the earliest, vast Archean Eon, there were few forms of life on Earth. It was a time of volcanoes, mountain building, erosion of land masses, and condensation of volcanic gas (which formed the oceans). As the Earth evolved, so did its life forms. Only simple forms of plant life and soft-bodied animals lived in the sea during the Proterozoic Eon. At the beginning of the more recent Paleozoic Era, 570 million years ago, shell-bearing animals first appeared in the sea.

Fish first appeared in the Silurian Period. They developed rapidly during the Devonian Period and flourish today. Amphibians appeared in the Devonian Period, followed by reptiles in the Pennsylvanian Period. The Mesozoic Era was the "age of dinosaurs," which became extinct at the end of the Cretaceous Period. Small mammals and birds appeared in the Jurassic Period, long before the dino-

Geologic time scale

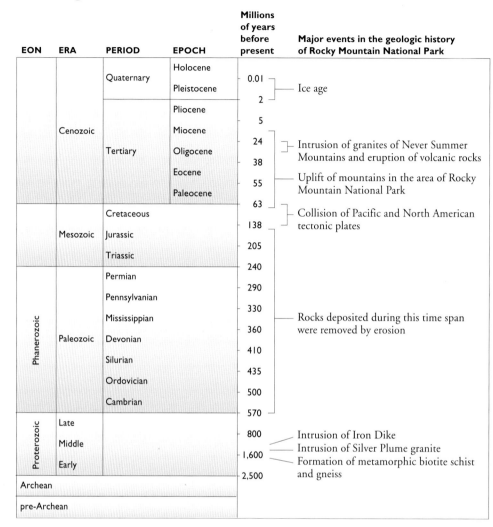

EON	ERA	PERIOD	EPOCH	Millions of years before present	Major events in the geologic history of Rocky Mountain National Park
Phanerozoic	Cenozoic	Quaternary	Holocene	0.01	Ice age
			Pleistocene	2	
		Tertiary	Pliocene	5	
			Miocene	24	Intrusion of granites of Never Summer Mountains and eruption of volcanic rocks
			Oligocene	38	
			Eocene	55	Uplift of mountains in the area of Rocky Mountain National Park
			Paleocene	63	
	Mesozoic	Cretaceous		138	Collision of Pacific and North American tectonic plates
		Jurassic		205	
		Triassic		240	
	Paleozoic	Permian		290	
		Pennsylvanian		330	
		Mississippian		360	Rocks deposited during this time span were removed by erosion
		Devonian		410	
		Silurian		435	
		Ordovician		500	
		Cambrian		570	
Proterozoic	Late			800	Intrusion of Iron Dike
	Middle			1,600	Intrusion of Silver Plume granite
	Early			2,500	Formation of metamorphic biotite schist and gneiss
Archean					
pre-Archean					

Geologic time and events reduced to a calendar year

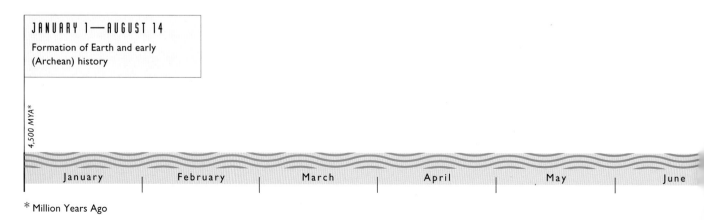

JANUARY 1—AUGUST 14

Formation of Earth and early (Archean) history

4,500 MYA*

January   February   March   April   May   June

* Million Years Ago

saurs died out. Our human ancestors appeared much later—a mere 3.5 million years ago.

Using the geologic time scale, it is easy to discuss events that happened over hundreds of millions of years. It is considerably more difficult to grasp the immense amounts of time involved. To put things into perspective, imagine that all of geologic time, from the formation of Earth to the present, took place during one calendar year. When in that year would various geological events have happened? How would fairly recent events fit into this time scale? The time line offers some perspective. ■

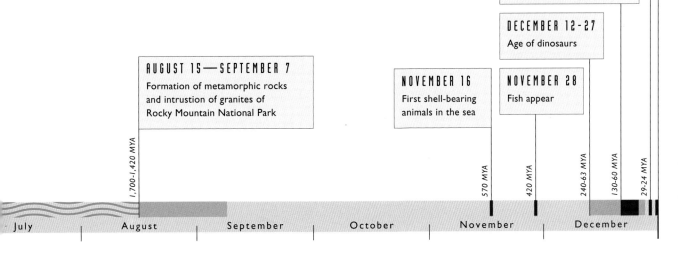

DECEMBER 31

5:00 p.m.
Ancestors of man   *3.5 MYA*

11:58 p.m.
Last Ice Age   *20,000 years ago*

11:59:46 p.m.
Birth of Christ   *2,000 years ago*

11:59:58 p.m.
(last 2 seconds)
Modern technology   *300 years ago*

DECEMBER 29

Intrusion of granites of
Never Summer Mountains
and eruption of volcanic rocks

DECEMBER 20-25

Collision of crustal plates that
form the mountains of Rocky
Mountain National Park

DECEMBER 12-27

Age of dinosaurs

AUGUST 15—SEPTEMBER 7

Formation of metamorphic rocks
and intrusion of granites of
Rocky Mountain National Park

NOVEMBER 16

First shell-bearing
animals in the sea

NOVEMBER 28

Fish appear

1,700-1,420 MYA

570 MYA

420 MYA

240-63 MYA

130-60 MYA

29-24 MYA

July   August   September   October   November   December

Ages of rocks in the park

From a geological point of view, some of the rocks in this park are very old. Others have formed relatively recently. Geologists are able to tell the **relative age** of different kinds of rocks by the way they are in contact with one another. The rocks' **absolute age** in years can be determined by the presence of radioactive minerals in some of the rocks.

RELATIVE AGE OF ROCKS

Determining the relative ages of different rock types by mapping their contact relationships is fairly simple. The diagram on the next page illustrates the relative age relationships of five common rock types in Rocky Mountain National Park. The biotite gneiss (pronounced NICE) and schist (1) contains layering and folds that end where they contact granite (2). The granite invaded and cut across the gneiss and schist layers. This indicates that the granite is the younger of the two rock types. In the same area, pegmatite dikes (3) have intruded both the metamorphic rocks and the granite, proving that these are younger still. After a period of erosion that planed a relatively smooth surface on the top of rocks (1), (2), and (3), volcanic rocks (4) were injected up through the gneiss and schist (1) and deposited on top of the older rocks. The volcanic rocks are the youngest of the four rock types. Most recently, the glacial moraines and tills (5) were deposited atop the other rock types, making them among the youngest geological features in the park.

ABSOLUTE AGE OF ROCKS

The absolute age of some rocks can be determined in years by using nature's own atomic clocks. Some metamorphic rocks and most igneous rocks contain minerals that have radioactive elements. Radioactive elements are converted into other elements by a process called radioactive decay. Each radioactive element decays at its own fixed rate. By measuring this rate of decay, geologists can use radioactive elements as atomic clocks.

For example, one radioactive element that can be used as an atomic clock is uranium. A certain form of uranium, isotope 235, decays to form a particular type of lead, isotope 207. Uranium 235 decays so slowly that it takes 713 million years for half of the uranium to change into lead 207. This rate of decay is called the half-life of the element. By measuring the amounts of uranium 235 and lead 207 in a rock sample, it is possible to calculate back to the time when there was only uranium 235—and no lead 207—in the rock. This reveals the rock's age. Several other radioactive elements, each with different half-lives, also can be used as atomic clocks.

The absolute ages of the major types of rocks in the park, along with their descriptions, are listed in the back of the book (page 71). ■

Relative age relationships of common rock types

(See discussion of Relative Age of Rocks on page 6)

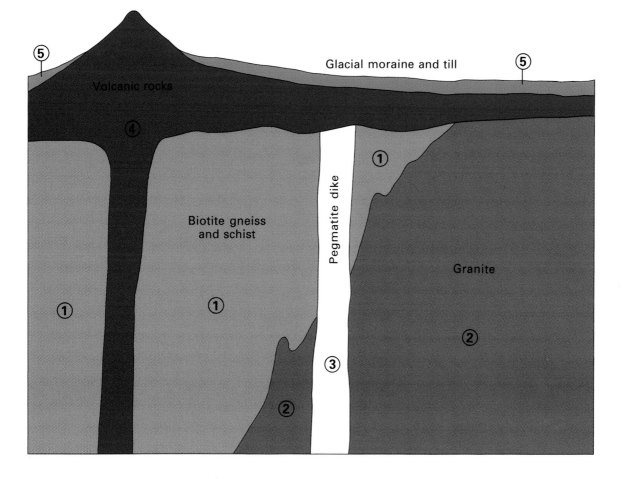

Generalized geologic cross section of Rocky Mountain National Park

This simplified geologic cross section traverses the park in a line that generally follows Trail Ridge Road in a northwest-southeast direction. The section portrays the rock types along the route—excluding glacial deposits—and shows variations in elevation along and near the road. ■

 Tertiary volcanics

 Tertiary granites

Middle Proterozoic (Silver Plume) granite and pegmatite

Early Proterozoic metamorphic rocks

 Fault

 Major fault zone

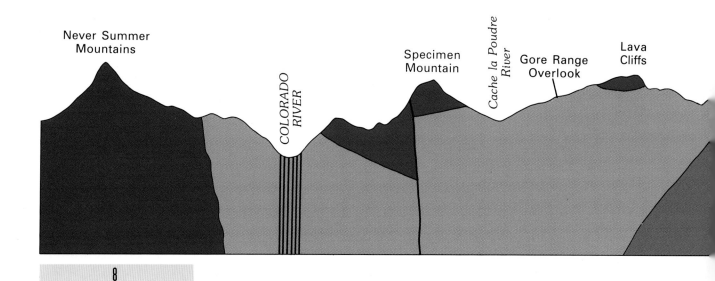

Brief geologic history

Most of the rocks in Rocky Mountain National Park—excluding the newer rocks of the Never Summer Mountains—originally were shale, siltstone, and sandstone, along with some volcanic rocks deposited about 1.8 to 2 billion years ago in an ancient sea (see the geologic time scale on page 4). Between 1.7 and 1.6 billion years ago, these sedimentary rocks were caught in a collision zone between sections of the Earth's crust called tectonic plates.

These rocks, then in the core of an ancient Proterozoic mountain range, were recrystallized into metamorphic rocks by enormous heat and pressure resulting from the collision. The shale, which contained mostly clay minerals and some very fine sand and silt, was converted into biotite schist. The layers with more sandstone were converted into biotite gneiss. These rocks are described and illustrated in this Trail Ridge Road tour.

Granites found in the park probably resulted from the melting of preexisting sedimentary or metamorphic rocks in the primordial crust shortly after the formation of the Earth. The Silver Plume granite that occurs in much of the east side of the park intruded upward into the metamorphic rocks about 300 million years after the formation of the Proterozoic mountains. We do not know what caused this igneous episode.

The high mountains that formed here during Proterozoic time were slowly eroded and reduced to a fairly flat surface, exposing the core of metamorphic rocks and granite. This erosion occurred over a long period, from approximately 1,300 million to 500 million years ago. Little else is known about the geologic events in this area during this time span because no rocks of that age are present in the region.

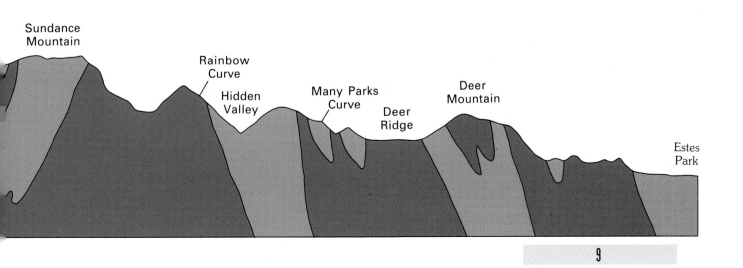

About 500 million years ago, this relatively flat area became covered with shallow seas. Over the next 200 million years, several hundreds of thousands of feet of Paleozoic sedimentary rocks were deposited on the old Proterozoic surface. During the middle Pennsylvanian Period, yet another mountain range was uplifted in this area. From it the Paleozoic Period sediments were eroded.

Sediments shed from these ancestral Rocky Mountains were deposited along the mountain flanks. Today, these make up the Fountain Formation, which comprises the Flatirons west of Boulder, Colorado; the spectacular walls of Red Rocks Park west of Denver; and the red rocks in the Garden of the Gods near Colorado Springs. The Fountain Formation appears as a fairly continuous outcrop along the east edge of the present mountain front.

The area that is now Rocky Mountain National Park was eroded again and intermittently covered by seas from the middle of the Permian Period to the end of the Cretaceous Period about 65 million years ago. Abundant bones and tracks found in sedimentary rocks in the Rocky Mountain region date back to Jurassic and Cretaceous times, indicating that dinosaurs lived here during those periods.

Major tectonic plates of the Earth's crust began to collide along what was then the western edge of North America about 130 million years ago. Uplift caused by this collision began to affect the area of the present Colorado Rockies about 70 million years ago. As the region began to rise, the Cretaceous sea withdrew and the thick layer of sedimentary rocks that had accumulated began to erode. Within a few million years, the sedimentary rocks of the Front Range had eroded away, and the Proterozoic igneous and metamorphic rocks again were exposed to erosion.

As uplift proceeded, deep fault zones formed as enormous stresses pulled the Earth's crust apart at what is now the west side of the park. This allowed granitic magmas to rise into the present area of the Never Summer Mountains. Between 29 and 24 million years ago, the magmas reached the surface and erupted as volcanoes. The tops of the volcanoes stood several thousand feet above the present granitic masses of the Never Summers, which since have been eroded to their present size. Lava flows and extensive ash beds from the volcanoes are preserved in several areas within the park.

From the plate collision to the present, rivers and streams have eroded the mountains and transported enormous amounts of sediment to the oceans. By the end of the Tertiary Period, the mountains of Rocky National Park were still fairly high but rounded. The area also was characterized by wide, V-shaped stream valleys.

Then, the Rocky Mountain National Park area saw more drama. About 2 million years ago, Earth's climate cooled and the Ice Age began. Large ice sheets ebbed and flowed across much of the Northern Hemisphere. During several major periods of glaciation—as well as several minor episodes—ice covered much of North America and Europe. The high mountain valleys filled with glaciers.

Rocky Mountain National Park felt the effects of the Ice Age. Glaciation in the park probably started about 1.6 million years ago. Specific evidence of the earliest glaciations doesn't exist because moraines formed by the early glaciers were destroyed by glaciers that followed later. Each time glaciers flowed down the mountain valleys they eroded the valley sides and bottoms, helping to straighten and deepen them, removing evidence of earlier glaciations.

Evidence of the last two major periods of ice accumulation is quite clear, however. The first of these two glacial periods is called the Bull Lake Glaciation. The Bull Lake advance began about 300,000 years ago and ended about 130,000 years ago. A few isolated remnants of moraines from the Bull Lake glaciers can be identified at various places in the park. They indicate that the amount of ice in the valleys then was equal to or greater than ice volume during the most recent period of glaciation.

After the Bull Lake glaciation came a warmer period that lasted about 100,000 years. The last major glacial episode, called the Pinedale Glaciation, began about 30,000 years ago when Earth's climate once again cooled. The Pinedale glaciers reached their maximum extent between 23,500 and 21,000 years ago. Most of the major valleys in the park were filled with glaciers during this time. One of the largest of the park glaciers, with a length of 13 miles (21 km), was in Forest Canyon just south of the high point of today's Trail Ridge Road. The largest glacier, about 20 miles (52 km) long, was the ice flow that occupied the Colorado River Valley on the west side of the park. The ice in many of these glaciers reached thicknesses of 1,000 to 1,500 feet (305 to 457 m).

Between 15,000 and 12,000 years ago, the climate warmed and the glaciers rapidly disappeared. The only glaciers found in the park today occupy locations that receive a large amount of snow blown across the mountain ridges into northeast-facing, shaded cirques where snow melts slowly during summer. None of these glaciers are remnants of Ice Age glaciers.

Some scientists believe that we are living today in a warming interglacial period. But they speculate that climates might cool again, and the glaciers could return.

The grandeur of Rocky Mountain National Park is the culmination of many geologic events: the formation of the rocks through hundreds of millions of years, the repeated uplift of the mountains by gigantic tectonic forces, and millions of years of erosion by water and ice that sculpted the mountains into their present forms. ■

O. B. Raup, W. A. Braddock, and R. F. Madole

Some notes about the tour

The self-guided geologic tour along Trail Ridge Road includes seventeen stops selected to introduce park visitors to the geology of Rocky Mountain National Park. This tour takes about 5.5 hours, assuming 15 minutes per stop and 1.5 hours of driving time. Distance between the Beaver Meadows Entrance Station and the last stop on the west side is 40.8 miles (65.6 km). The distance is 41.6 miles (66.9 km) if you begin at the Fall River Entrance. Mileage is given between stops, and a cumulative mileage is given from each entrance station. Car odometers can vary slightly, so please be on the lookout for the stops as you travel over the road.

Several of the stops on this tour are at parking areas marked by the numbered arrowhead-shaped signs that complement the Trail Ridge Road guide pamphlet. When applicable, these numbers are included in this text to help locate the stops.

Travelers starting at either the Beaver Meadows or Fall River entrance should follow **WESTBOUND** directions in the upper corner of the following pages. Travelers starting at the Grand Lake Entrance Station should start on page 64 of the book and follow **EASTBOUND** directions in the upper corner of the pages. To calculate point-to-point distance, use your odometer or the trip mileage recorder, if your vehicle has one.

Discussions at some of the stops are somewhat repetitious. This is so visitors traveling in different directions receive the same information.

A foldout geologic map of the area along Trail Ridge Road is included at the end of the book. This oblique view map depicts the landscape and the distribution of the various rock types. Referring to the map at the tour stops will help you orient yourself. You will find binoculars helpful for viewing features that are far from the road and watching the park's abundant wildlife.

Some sections of Trail Ridge Road are steep and narrow, so be cautious of oncoming cars while you are looking at the rocks in roadcuts. **Please remember that collecting specimens of rocks, minerals, or plants is prohibited.**

Have a safe and enjoyable trip.

BEAVER MEADOWS STARTING POINT (WESTBOUND)

The Beaver Meadows Entrance Station is on U.S. Highway 36 southwest of Estes Park. Set your trip odometer to 0 or record your car mileage here. Proceed 2.4 miles (3.9 km) west to Beaver Meadows Overlook and turn to page 15.

The Fall River Entrance Station is on U.S. Highway 34 northwest of Estes Park. Set your trip odometer to 0 or record your car mileage here. Proceed west 1.7 miles (2.7 km) to the Sheep Lakes parking area on the northwest side of Horseshoe Park and turn to page 17.

Trail Ridge Road, Rocky Mountain National Park

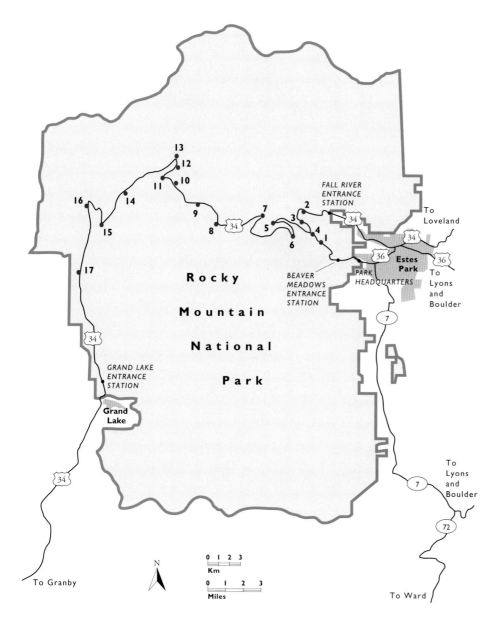

1. Beaver Meadows Overlook
2. Sheep Lakes
3. Horseshoe Park Overlook
4. Deer Ridge Junction
5. Iron Dike
6. Many Parks Curve
7. Rainbow Curve
8. Forest Canyon Overlook
9. Rock Cut
10. Lava Cliffs
11. Gore Range Overlook
12. Alpine Visitor Center
13. Medicine Bow Curve
14. Milner Pass
15. Farview Curve
16. Rock Outcrop
17. Bowen-Baker Trailhead

Rocky Mountain elk

GRAND LAKE STARTING POINT (EASTBOUND)

The Grand Lake Entrance Station is on U.S. Highway 34, 1.5 miles (2.4 km) north of the highway junction into Grand Lake, a town located on the west side of Rocky Mountain National Park. Set your trip odometer to 0 or record your car mileage here. Proceed 6.3 miles (10.1 km) north to the Bowen-Baker Trailhead and turn to page 64. ■

Beaver Meadows Overlook

GEOLOGY 1 STOP

The parking area is next to a house-size rock formation on the left side for westbound travelers, on the right side for eastbound travelers.

WESTBOUND

- Distance from Beaver Meadows Entrance Station to this stop
  2.4 miles (3.9 km)

- Next stop:
  Deer Ridge Junction
  0.7 mile (1.1 km)

EASTBOUND

- Distance from Grand Lake Entrance Station to this stop
  37.3 miles (60 km)

- Distance to Beaver Meadows Entrance Station
  2.4 miles (3.9 km)

This completes your geology tour of Rocky Mountain National Park.

This stop affords a view to the south across Beaver Meadows toward Longs Peak (see photograph below). At 14,255 feet (4,345 m) above sea level, flat-topped Longs Peak, the tallest mountain on the skyline in front of you, is the highest summit in Rocky Mountain National Park. The mountain is composed almost entirely of an igneous rock called Silver Plume granite.

This massive rock is relatively hard and resistant to erosion. The characteristics and origin of Silver Plume granite are discussed at Rainbow Curve (page 29).

Some rounded boulders and outcrops of Silver Plume granite can be found just off the edge of the parking area. These rocks have been exposed to weathering for thousands of years, which explains their coarse, rough tex-

Longs Peak, the jewel in the crown of Rocky Mountain National Park, rises above Moraine Park and Beaver Meadows. The two prominent moraines in the middle foreground were deposited by the Big Thompson glacier.

Weathering has etched depressions in Silver Plume granite.

ture. Their gritty surface is composed of grains of quartz and feldspar loosened by the removal of more easily weathered crystals of mica. The granite here is more deeply weathered than the granite at higher elevations because it has been exposed longer. The warmer conditions here also cause more rapid chemical decomposition than occurs in the colder, higher regions.

Dish-shaped depressions in the granite were formed by weathering processes and the freezing and thawing of water that collects in minor recesses on the rock surfaces. Water works its way along crystal boundaries, freezing loosens the crystals, and the wind blows the crystals out of the depressions.

The two low, tree-covered horizontal ridges just beyond the meadows are moraines deposited along the margins of the ancient Big Thompson glacier, which once flowed out of the mountains to the right. Moraines are composed of debris scoured by glaciers from the sides of the high mountain valleys. These moraines are discussed in more detail at Many Parks Curve (page 25).

Westbound travelers proceed 0.7 mile (1.1 km) to Deer Ridge Junction and turn to page 22.

This stop completes the tour for Eastbound travelers headed for the Beaver Meadows Entrance Station. ∎

Sheep Lakes

GEOLOGY 2 STOP

This stop is at a large parking area on the left side of the road for westbound travelers, on the right side for eastbound travelers. The stop is next to two small lakes called Sheep Lakes (see fold-out map). Two exhibits at this parking area describe some of the features seen from here.

To the west, the valley you see across Sheep Lakes has a U-shaped profile caused by erosion from the Fall River glacier as it flowed out of the mountains (see photograph below). This U is the characteristic shape of a glacial valley. Valleys that are cut exclusively by running water instead of ice have a V-shaped profile, as explained in more detail at the Forest Canyon Overlook (page 33).

As the Fall River glacier flowed in this direction, it plucked rock debris from the sides of the valley and deposited the material at this lower level. The mix of rock and sediment was left in

WESTBOUND

- Distance from Fall River Entrance Station to this stop
  1.7 miles (2.7 km)

- Next stop: Horseshoe Park Overlook
  1.4 miles (2.2 km)

EASTBOUND

- Distance from Grand Lake Entrance Station to this stop
  39.9 miles (64.2 km)

- Distance from Horseshoe Park Overlook
  1.4 miles (2.2 km)

- Distance to Fall River Entrance Station
  1.7 miles (2.7 km)

This stop completes your geology tour of the park.

HIDDEN VALLEY — RAINBOW CURVE — SUNDANCE MOUNTAIN — FALL RIVER VALLEY — SOUTH LATERAL MORAINE

The U-shaped Fall River Valley was shaped by a glacier that flowed from the high mountains more than 15,000 years ago. The south lateral moraine blocks the view of Hidden Valley.

17

the form of narrow ridges called moraines. The south lateral—or side—moraine of the Fall River glacier is visible between the valley floor and the switchbacks of Trail Ridge Road. This moraine hides the valley behind it, which is named Hidden Valley (see photograph on page 17). The terminal—or end—moraine is to the east (left), marked by the low hills that cross the valley. The north lateral moraine of the Fall River glacier is on the other side of the road from this parking area. These glacial deposits are composed of unsorted materials that range in size from sand to large boulders.

The flat valley floor seen from here resulted from sediments deposited in a lake that was dammed by the terminal moraine of the Fall River glacier. After a time, the lake filled with sediment, leaving the present valley bottom.

During the last retreat of the glaciers between 15,000 and 12,000 years ago, some large blocks of ice were left imbedded in the lake sediments as the glacier melted back. When these blocks melted, the depressions left in the lake sediments filled with water, forming small lakes. These particular lakes have been named Sheep Lakes because they are favorite mineral licks for bighorn sheep.

This stop completes the tour for Eastbound travelers headed for the Fall River Entrance Station. ■

Bighorn sheep

Horseshoe Park Overlook

WESTBOUND

- Distance from Fall River Entrance Station to this stop
  3.1 miles (5 km)

- Distance from Sheep Lakes
  1.4 miles (2.2 km)

- Next stop: Deer Ridge Junction
  0.8 mile (1.3 km)

EASTBOUND

- Distance from Grand Lake Entrance Station to this stop
  38.5 miles (61.9 km)

- Distance from Deer Ridge Junction
  0.8 mile (1.3 km)

- Next stop: Sheep Lakes
  1.4 miles (2.2 km)
  Turn to page 17

GEOLOGY 3 STOP

This stop is at a parking area on the left side of the road for westbound travelers, on the right side for eastbound travelers. It is located near a curve up the hill from the floor of Horseshoe Park. There are two entrances to this parking area. An exhibit here shows a sketch of the area to the north and names the mountains.

The highest peak on the northwest skyline is Ypsilon Mountain. The mountain was named this because its narrow, snow-filled gullies suggest the shape of the Greek letter Y, usually spelled upsilon.

Ypsilon Mountain is composed primarily of metamorphic rocks that from this distance appear to have horizontal stripes. These rocks once were layered sedimentary rocks. But they were recrystallized hundreds of millions of years ago, deep in the Earth, by great heat and pressure during ancient mountain-building processes. The rocks under your feet in this parking

Ypsilon Mountain dominates the skyline of the Mummy Range. The alluvial fan seen here on the far side of the valley was deposited by floodwaters when Lawn Lake Dam failed in 1982.

19

BIGHORN MOUNTAIN

McGREGOR MOUNTAIN

NORTH LATERAL MORAINE

The Fall River meanders along the valley floor beneath Bighorn and McGregor mountains.

This telephoto view shows the fan shape of the alluvial deposit from the Lawn Lake flood. The photograph was taken 0.45 mile (0.73 km) north of Many Parks Curve.

area are some of these metamorphic rocks. The metamorphic rocks of the park are discussed in greater detail at Rock Cut (page 38).

On the far side of the valley floor, this side of Ypsilon Mountain, is a fan-shaped deposit of sand, gravel, and boulders, some as large as a car. These materials were dumped by floodwaters after a dam broke near the upper end of the Roaring River Valley in the early 1980s. Look for another good view of this sandy deposit, called an alluvial fan, from Rainbow Curve, where the flood

A close-up look at the meandering Fall River.

is discussed in more detail (page 29).

As the Fall River flows across the flat lake sediments in the valley floor it meanders, looping back and forth. When a river flows down a steeper valley, the energy of the water is directed to eroding the valley floor. But when water reaches a flat valley it is no longer able to cut downward with the same force. The energy of the moving water is deflected sideways, first in one direction, then another. As the stream moves its channel sideways, its meandering loops become more and more convoluted. Eventually, parts of the channels run into one another. If one of these loops is cut off, the stream will leave behind isolated sections that can be shaped like horseshoes (see photograph above). This is how Horseshoe Park earned its name.

Straight across the valley is Bighorn Mountain. To the right is McGregor Mountain. Both of these mountains have exposed knobs of granite, which weather into fairly smooth domes. An especially good example of one of these domes is on the west side of McGregor Mountain (see photograph on page 20). These domes are discussed in more detail at Many Parks Curve (page 25). ■

WESTBOUND

- Set your trip odometer to 0 or record a new starting mileage at this stop.
- Next stop: Iron Dike
  2.7 miles (4.3 km)

EASTBOUND

- Distance from Grand Lake Entrance Station to this stop
  37.7 miles (60.6 km)

- Distance from Iron Dike
  2.7 miles (4.3 km)

- Next stop: Beaver Meadows Overlook
  0.7 mile (1.1 km)
  toward Beaver Meadows Entrance Station Turn to page 15

  or Horseshoe Park Overlook
  0.8 mile (1.3 km)
  toward Fall River Entrance Station Turn to page 19

Deer Ridge Junction

GEOLOGY 4 STOP

Deer Ridge Junction is on the divide between Moraine Park to the south and Horseshoe Park to the north (see fold-out map). It marks the beginning of Trail Ridge Road on the east side of the park.

Westbound travelers should set their trip odometer to 0 or record their starting mileage here. Proceed 2.5 miles (4 km) west (uphill to the right) to the turnoff into Hidden Valley and the Iron Dike.

Eastbound travelers have a choice of going either toward the Beaver Meadows Entrance Station or the Fall River Entrance Station (see fold-out map in the back of this book and general location map on page 13).

Between Deer Ridge Junction and Hidden Valley is a series of beaver ponds along Hidden Valley Creek. The ridge on the other side of the ponds is the south lateral moraine deposited by the Fall River glacier. This is the moraine that hides Hidden Valley from Horseshoe Park. ∎

Majestic Longs Peak dominates the view south of Deer Ridge Junction.

Iron Dike

GEOLOGY
5
STOP

Westbound travelers turn right at a sharp curve in the road into the parking lot of the former Hidden Valley ski area (Arrowhead Sign 2) and drive 0.2 mile (0.3 km).

Eastbound travelers drive 1.4 miles (2.2 km) down the road from the lower parking area of Many Parks Curve to a sharp curve in the main road, turn left into the parking lot of the former Hidden Valley ski area (Arrowhead Sign 2), and drive 0.2 mile (0.3 km).

The Iron Dike is on the right side of the valley (northwest side) about a third of the way into the parking lot. The dike is approximately 70 feet (21 m) wide at this location and looks like horizontal rows of rust-colored stairs up the hillside (see photograph below).

A dike is a wall-like intrusion of igneous rock that cuts across the surrounding rock. The dike at this site is part of a series of dikes that extends from southeast to northwest through the park. The dikes are shown on the fold-out map in the back of this book. These dikes intruded into the ancient

WESTBOUND

- Distance from Deer Ridge Junction to the turnoff into this stop
  2.5 miles (4 km)

- Next stop:
  Many Parks Curve
  1.7 miles (2.7 km)
  to the upper parking area

EASTBOUND

- Distance from Grand Lake Entrance Station to the turnoff into this stop
  33.4 miles (53.7 km)

- Distance from the Many Parks Curve lower parking area to the turnoff
  1.6 miles (2.6 km)

- Distance from the turnoff to the Iron Dike
  0.2 mile (0.3 km)

- Next stop:
  Deer Ridge Junction
  2.7 miles (4.3 km)

Turn to page 22

Red-hot magma intruded granite and metamorphic rocks to form a giant gabbro dike about 1.3 billion years ago. The result is this stair-step outcropping.

metamorphic rocks and granites about 1.3 billion years ago, making them about 100 million years younger than the Silver Plume granite discussed at Rainbow Curve (page 29).

Although dikes can be of various compositions, these so-called iron dikes are composed of a rock called gabbro, which contains the minerals plagioclase, augite, and magnetite. Some of these minerals contain the element iron, which when exposed to weathering combines with oxygen to form iron oxide minerals such as limonite, a major component of rust. This gives the dike rocks a rusty brown appearance.

The dikes were liquid when they intruded the other rocks. Under enormous pressure from great depths in the Earth's crust, this liquid forced its way upward through cracks in the rocks and shoved the other rocks aside. After the liquid was in place, it began to cool. The crystals near the center of the dike are large because they had a long time to grow during slow cooling. The crys-tals next to the border cooled rapidly and are small. As the entire dike cooled, shrinkage cracks formed at right angles to the sides of the dike. These cracks, or joints, give the dike its ledgy, stair-step appearance.

One segment of the Iron Dike crosses Trail Ridge Road between Many Parks Curve and Rainbow Curve. Unfortunately, there are no convenient parking areas nearby. Other segments of the Iron Dike are visible on the south face of Mount Chapin and on the upper southeast slope of Sundance Mountain (see fold-out map). However, these are difficult to see in less-than-ideal light conditions.

Proceed on the tour by going back to Trail Ridge Road.

Westbound travelers turn right and drive up the hill to the upper parking area of Many Parks Curve.

Eastbound travelers turn left and go down the road to Deer Ridge Junction. ■

Many Parks Curve

WESTBOUND

- Distance from Deer Ridge Junction to this stop (upper parking area)
  4.4 miles (7.1 km)

- Distance from Iron Dike
  1.7 miles (2.7 km)

- Next stop: Rainbow Curve
  4 miles (6.4 km)

EASTBOUND

- Distance from Grand Lake Entrance Station to this stop
  33.4 miles (53.7 km)

- Distance from Rainbow Curve (lower parking area)
  4.1 miles (6.6 km)

- Next stop: Iron Dike
  1.6 miles (2.6 km)

Turn to page 23

GEOLOGY 6 STOP

Westbound travelers must continue around the curve and park in the upper parking area on the right side of the road. Eastbound travelers may park in either the upper parking area on the left side of the road or in the lower parking area around the curve on the right side of the road. This stop is at Arrowhead Sign 3.

At this stop on Many Parks Curve, visitors can find good views from both the boardwalk next to the road between the two parking areas and the lower parking area. Longs Peak is easily seen from the boardwalk; it is the prominent, flat-topped mountain on the skyline to the south (see photograph below). Its football field-sized summit is a remnant of the high, rolling terrain that is dominant along much of the upper part of Trail Ridge Road. This gently rolling surface is discussed in more detail at Forest Canyon Overlook (page 33).

During the Ice Age, when most of the Northern Hemisphere was in the

Moraines left behind by the Big Thompson glacier frame Moraine Park below Longs Peak. The rounded knobs protruding from the trees are granite domes.

A picturesque tree grows from an outcrop of metamorphic rocks containing lenses of pegmatite. These rocks were altered by heat and pressure during the formation of an ancient mountain range about 1.7 billion years ago.

Vertical layers of biotite schist enclose light-hued layers of pegmatite.

McGREGOR MOUNTAIN

LUMPY RIDGE

The rounded knobs on McGregor Mountain and Lumpy Ridge are exfoliation domes of Silver Plume granite.

grip of cold, wet climates, glaciers flowed out of the high mountains you see to the right. Two large moraines deposited along the sides of one of these large glaciers appear as low, horizontal, tree-covered ridges in the valley to the left of Longs Peak (see photograph on page 25). These moraines were deposited 20,000 years ago by the glacier that flowed out of the mountains down the present valley of the Big Thompson River. The moraines are composed of material that the glacier and its tributaries scraped off the sides of the valleys farther upstream. The glaciers deposited the material as they moved out into the open valley. Glaciers are dis-

cussed in more detail at Forest Canyon Overlook (page 33).

The rocky knob next to the lower parking area is a complex mixture of rock types. These rocks are composed of biotite schist, Silver Plume granite, and pegmatite. The shiny black mineral in these rocks is biotite. The layers in the biotite schist originally were sedimentary rocks formed by layers of sand, silt, and clay. These rocks are discussed in more detail at Rainbow Curve (page 29) and the Rock Cut (page 38).

The rounded knobs on McGregor Mountain and Lumpy Ridge (see photograph above) are smooth masses of granite with a very uniform structure.

This exfoliation dome adorns the side of McGregor Mountain. The photograph was taken 0.7 mile (1.13 km) west of the Fall River Entrance Station.

These knobs, called exfoliation domes, are formed as erosion of overlying rocks releases pressure on the granite, allowing the granite to expand. As it expands, it cracks into curved layers somewhat like the layers of an onion. Smooth domes are left after the curved layers break away and the fragments are removed by erosion. Similar exfoliation domes are common in other areas of massive granite deposits, notably Yosemite National Park in California. Other characteristics of the granite are discussed at Rainbow Curve (page 29). ■

Rainbow Curve

GEOLOGY 7 STOP

This stop is at a large parking area on the outside of a wide curve in the road on the right for westbound travelers, on the left for eastbound travelers (Arrowhead Sign 4).

WESTBOUND

- Distance from Deer Ridge Junction to this stop
  8.4 miles (13.5 km)

- Distance from Many Parks Curve (upper parking area)
  4 miles (6.4 km)

- Next stop: Forest Canyon Overlook
  2.9 miles (4.7 km)

EASTBOUND

- Distance from Grand Lake Entrance Station to this stop
  29.3 miles (47.1 km)

- Distance from Forest Canyon Overlook
  2.9 miles (4.7 km)

- Next stop: Many Parks Curve
  4.1 miles (6.8 km)
  to lower parking area
  Turn to page 25

Rainbow Curve offers visitors magnificent views of the east side of the park. Straight ahead to the east is the valley of the Fall River, which meanders through the flat floor of Horseshoe Park. On the left side of the valley is an alluvial fan deposited by floodwaters that raged down the Roaring River when a dam collapsed on an otherwise perfect summer morning, July 15, 1982.

The Lawn Lake Dam broke around 5:30 a.m., and the water that had been held in the reservoir roared down the Roaring River Valley with the sound and fury of a runaway freight train. Within minutes, the water's tremendous force scoured sand, gravel, and boulders from the Roaring River Valley and dumped the material at the edge of the broad Fall River Valley. The floodwaters quickly spread out across

From Rainbow Curve, visitors see this grand vista sweeping east down the Fall River Valley. At left is the alluvial fan deposited by Lawn Lake floodwaters. Exfoliation domes of granite rise from Bighorn Mountain, McGregor Mountain, and Lumpy Ridge. The south lateral moraine of the Fall River glacier blocks the entrance to Hidden Valley.

The view toward Mummy Mountain shows the upper boundary of tree growth, called the treeline, which is nearly level in this area.

MUMMY MOUNTAIN

TREELINE

The rounded outcrop across the road is a typical example of Silver Plume granite in the park.

this broad valley, spilled down its course to the east, and flooded downtown Estes Park.

The newly formed alluvial fan dammed the Fall River and created a new lake, which you can see on this side of the fan. This lake will be there until it fills with sediment, or until its over-

flow cuts far enough into the alluvial fan to allow the lake to empty.

Steeply downhill to the east is a series of beaver ponds between the lower part of Trail Ridge Road and a ridge to the left (see photograph on page 29). The ridge is a lateral moraine deposited by the glacier that flowed down the

Close inspection of the granite reveals crystals of quartz, feldspar, and mica.

A closer view of the rock's polished surface shows interlocking crystals of quartz (glassy gray), feldspar (white), and mica (black).

Fall River Valley from the high country to the west. Hidden Valley was named because it is hidden from Horseshoe Park by this moraine.

The rocks around this parking area are Silver Plume granite. The large outcrop on the other side of the road, and the cut blocks used in construction of the rock wall around the parking lot, show the characteristics of this rock (see photographs on page 30 and above).

Granite is an igneous rock composed primarily of the minerals quartz, feldspar, and mica. The quartz is glassy and gray, and the feldspar is white or slightly pink. The mica is gray to black with shiny, highly reflective surfaces that make the rocks sparkle where they have been freshly broken. All of these minerals have about the same crystal size, and the crystals are locked into one another like a jigsaw puzzle (see photographs on page 31).

The crystals grew in this manner from a liquid that cooled very slowly. The original molten material invaded the older metamorphic rocks approximately 1.4 billion years ago when rock materials deep within the Earth's crust melted and were forced into the overlying core of an ancient mountain range. The relative ages of the granite and the metamorphic rocks are discussed at the Rock Cut (page 38).

The treeline (sometimes called timberline) is the upper limit of tree growth. Look for it just above this parking area. In the park, treeline is between 11,000 and 11,500 feet (3,353 and 3,505 m) above sea level. To the north, the treeline is almost a straight, level line (see photograph on page 30). The reasons for this growth limit are discussed in more detail at the Alpine Visitor Center (page 50).

Westbound travelers will drive through the treeline into the zone of tundra on the way to the Forest Canyon Overlook.

Eastbound travelers will quickly descend below treeline into the forest. ■

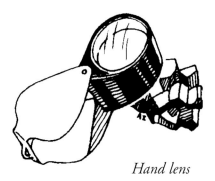

Hand lens

Forest Canyon Overlook

GEOLOGY 8 STOP

The parking area for this stop is on the south side of the road. It's on the left side for westbound travelers, the right side for eastbound travelers (Arrowhead Sign 5). Walk from the parking area to the viewing platform at more than 11,700 feet above sea level. You will be treated to a fantastic view of Forest Canyon and the glaciated mountains beyond.

WESTBOUND

- Distance from Deer Ridge Junction to this stop
  11.3 miles (18.2 km)
- Distance from Rainbow Curve
  2.9 miles (4.7 km)
- Next stop: Rock Cut
  2.1 miles (3.4 km)

EASTBOUND

- Distance from Grand Lake Entrance Station to this stop
  26.4 miles (42.5 km)
- Distance from Rock Cut
  2.1 miles (3.4 km)
- Next stop: Rainbow Curve
  2.9 miles (4.7 km)
  Turn to page 29

The landscape that you see here today resulted from a sequence of events spanning millions of years. About 130 million years ago, the Pacific tectonic plate began to collide with the western edge of the North American plate, and the western part of the continent began to be uplifted. About 70 million years ago, the effects of the collision and uplift reached the Rocky Mountains. The uplift continued periodically through most of the Tertiary Period. The mountains reached their present elevation about 6 million years ago (see

Glacier-carved Terra Tomah Mountain dominates the view across Forest Canyon toward the Continental Divide. The mountain's smooth upper surface is a remnant of the rolling upland terrain that characterized the area before the Ice Age.

Forest Canyon glaciation

The scene here as it may have appeared about 2.5 million years ago, before the mountains and valleys were sculpted by glaciers.

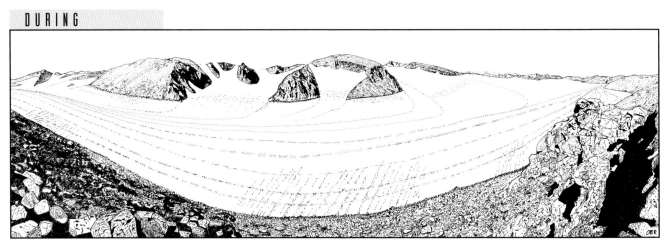

The same scene 20,000 years ago when the valleys were filled with glacial ice.

LONGS PEAK STONES PEAK TERRA TOMAH MTN. MT. IDA NEVER SUMMER MOUNTAINS

The vista as it appears today.

Looking southeast to Longs Peak, a telephoto camera shows the flat upper surface that was part of the pre-glacial topography. A huge glacial trough flanks the peak's summit on the right.

geologic time scale on page 4).

By the end of the uplift period, this area was a gentle, rolling terrain, much like the area on the upper part of Trail Ridge to the northwest. The tops of the mountains across Forest Canyon—Terra Tomah Mountain and Stones Peak—are remnants of the rolling upland, as are Flattop Mountain and the top of Longs Peak several miles to the east (see photograph above and illustrations on page 34).

While the land rose, streams flowing across the area acquired more downward erosion energy as the slope to the oceans increased and the water flowed faster. As water flows faster, it has more energy to carry sediments such as sand, gravel, and boulders. These stream-borne sediments cut the underlying bedrock like a fluid sawblade. Rapidly, the stream valleys became deep canyons.

The erosion by the streams was aided by weakness of the rocks caused by countless faults and cracks formed over the millennia during several episodes of mountain building. Forest Canyon is straight for many miles because the Big Thompson River follows a major fault zone in the rock (see foldout map at the back of this book).

About 2 million years ago, the climate over Earth's Northern Hemisphere became much cooler. Because this area had been uplifted almost to its present elevation, the combination of cooler climate and increased snowfall resulted in the accumulation of more snow than could melt. So, the valleys filled with snow and ice. When the ice in the valleys reached a thickness of about 200 feet (60 m), it started to flow slowly downhill under the force of gravity.

A body of ice becomes a glacier as

soon as it starts to move. The moving glacial ice plucked rocks from the valley walls. The mass acted as giant rasps and files to erode the floor and walls of the valley as the glacier moved slowly downward. Because of the weight of the ice, and because of its easily deformed (plastic) nature, the glacier cut downward over the width of the valley. In time, the glaciers changed the original V-shaped stream valleys into valleys with U-shaped glacial profiles. A prominent U-shaped valley is Glacier Gorge, immediately to the right of Longs Peak (see photograph on page 35).

Forest Canyon was cut deeper than the tributary valleys because the glacier in Forest Canyon was larger and had more cutting power than the smaller tributary glaciers. Since the bottoms of the tributary valleys were not cut as deeply as Forest Canyon, a steplike dropoff into the main valley was left behind when the ice melted between 15,000 and 12,000 years ago. The U-shaped valley remaining above the dropoff is called a hanging valley. The streams in the hanging valleys flow into the main valley as waterfalls and rapids.

The heads of the tributary valleys end in amphitheater-shaped cirques (pronounced SERKS). Cirques were created by the glaciers after snow and ice accumulated at the upper ends of the valleys. The glaciers continually enlarged the cirques and moved the headwalls back into the mountain ridges by plucking rocks from the steep cliffs and carrying them downstream. You can see where glaciers formed a small cirque on this side of TerraTomah Mountain (see photograph on page 33). Many of the high valley cirques contain beautiful lakes.

The ridge to the right of Mount Ida, west across Forest Canyon, exhibits the pronounced layering of the metamorphic rocks in this part of the park. These are discussed in greater detail at the Rock Cut (page 38).

Forest Canyon Overlook is about halfway between the west end of the canyon at Forest Canyon Pass and the east end of the canyon where it enters Moraine Park. The west end of the canyon can be seen from Gore Range Overlook (page 47).

Between Forest Canyon Overlook and Rock Cut, on the high rolling upland terrain, are solifluction (soil flow) terraces. These formed as water-saturated soil slowly moved downhill during countless daily and seasonal cycles of freezing and thawing. Most of this soil movement took place during periods of maximum glaciation, the last being about 20,000 years ago.

While the glaciers occupied the valleys, this upland area was a cold, snowy, windy place with deep permafrost. During the summer months, the ice in the upper few feet of the soil melted, and this upper layer became water saturated. This soft layer flowed slowly down even the most gentle slopes.

Rock streams and solifluction terraces cover the south slope of Sundance Mountain.

Photo by William C. Bradley

FOREST CANYON OVERLOOK

These solifluction terraces can be found between Forest Canyon Overlook and Rock Cut.

Photo by Douglass E. Owen

Night and day freezing and thawing helped move the soil downhill.

As parts of the soil moved, steps formed. Some steps held small ponds of water on the upper surface. Examples of these steps can be seen when the sun angle is low (see photographs above). Other periglacial features can be seen on this high rolling surface. Some additional features are described in the area between Rock Cut and Lava Cliffs. ■

WESTBOUND

- Distance from Deer
  Ridge Junction to
  this stop
      13.4 miles (21.6 km)

- Distance from Forest
  Canyon Overlook
      2.1 miles (3.4 km)

- Next stop: Lava Cliffs
      2 miles (3.2 km) west

EASTBOUND

- Distance from Grand
  Lake Entrance Station
  to this stop
      24.3 miles (39.1 km)

- Distance from
  Lava Cliffs
      2 miles (3.2 km)

- Next stop: Forest
  Canyon Overlook
      2.1 miles (3.4 km)

Turn to page 33

Rock Cut

GEOLOGY 9 STOP

The parking area is on the right for westbound travelers, on the right or left for eastbound travelers (Arrowhead Sign 6).

The metamorphic rocks in this roadcut originally were sedimentary rocks deposited in an ancient sea between 1.8 and 2 billion years ago. Layers of clay, silt, and sand were compressed into sedimentary rocks such as shale, siltstone, and sandstone. Around 1.7 billion years ago, these rocks were caught in a mountain-building plate collision that subjected them to enormous heat and pressure.

Over millions of years, the great heat and pressure caused the shale and silt to recrystallize into schist, and the interbedded layers of siltstone and sandstone became gneiss. The clay minerals became micas, both biotite and muscovite. In addition, some other

METAMORPHIC ROCKS

PEGMATITE

At Rock Cut, Trail Ridge Road slices through ancient metamorphic rocks and pegmatite veins.

Irregular layers and masses of pegmatite interweave layers of biotite schist here. Much of this pegmatite was formed when the biotite schist was partially melted during metamorphism.

complex silicate minerals were formed. The biotite schist contains a high percentage of biotite mica, which causes the rock to appear generally dark and shiny. The biotite gneiss contains mostly quartz and feldspar, with some biotite in indistinct layers.

Silver Plume granite, which occurs in irregular patches toward the east end of the road cut, was injected as liquid magma into the metamorphic rocks around 1.4 billion years ago. Farther to the east, these rocks become massive granite similar to the rocks at Rainbow Curve.

Another form of igneous rock that can be seen here is pegmatite. A conspicuous vein of pegmatite cuts through the metamorphic rocks in the main road cut at this stop (see photograph on page 38). This pegmatite is obviously younger than the metamorphic rocks because it cuts across them. Pegmatite

Large crystals of feldspar and quartz, paired with shiny masses of flaky mica, characterize the veins and pods of pegmatite.

in this part of the park is composed of very coarse crystals of quartz, feldspar, some light-colored mica and, occasionally, black tourmaline.

There are actually two stages of pegmatites in the rocks along Trail Ridge Road. The first formed during

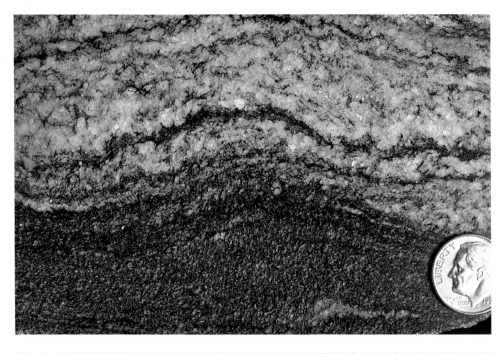

This close-up view displays a layer of biotite gneiss adjacent to layers of biotite schist.

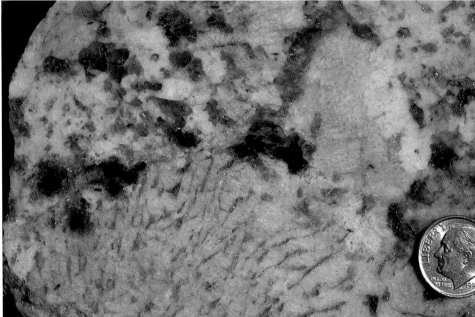

Polished pegmatite shows the coarse texture of the large interlocking crystals of quartz (glassy gray), feldspar (white and pink), and mica (black).

the metamorphism of the schist and gneiss when the heat and pressure were so great that some of the rock melted and migrated into the layers of the metamorphic rocks. These small layers and patches of pegmatite tend to run parallel to the layers of the metamorphic rocks.

The second stage of pegmatites formed during the late stages of the Silver Plume granite's formation and invasion into the metamorphic rocks. These later pegmatites tend to cut across the grain and texture of the metamorphic rocks and the granite itself. The two pegmatites are similar in appear-

The valley of Gorge Lakes, directly across Forest Canyon, was sculpted by a tributary glacier, which fed the main glacier in Forest Canyon. The lakes occupy several small basins scooped out of bedrock by the glacier. Snow veils most of the lakes in this photograph, but as many as six can been seen in late summer.

ance, but they can be distinguished from one another by their radioactive minerals (see the discussion on absolute age of rocks on page 6).

Across Forest Canyon from the Rock Cut stop is a tributary valley containing several lakes. These lakes occupy stair-step depressions scooped out by glacial ice as it moved toward the main glacier in Forest Canyon. Strings of similar lakes in high glaciated valleys of the Swiss and Italian Alps are called pater noster lakes. The name means "our father" in Latin, and alludes to

their resembling a string of rosary beads.

The Tundra World Nature Trail, which departs from this parking area, affords visitors the opportunity to walk through the high tundra. The tundra soil is very thin and fragile, **so please stay on the trail.**

The area between Forest Canyon Overlook to the east and the Lava Cliffs to the west is typical of the upland, rolling terrain that existed in much of Rocky Mountain National Park prior to glaciation. The relatively gentle terrain was dissected by some deep, and

Rocks thrust up to the ground surface through the countless cycles of freezing and thawing common in high tundra.

some not-so-deep, V-shaped valleys cut by streams. When the glaciers started accumulating about 2 million years ago, the ice was confined mostly to the valleys, while the higher elevations were normally free of snow.

This high surface was barren, snowy, windswept, and frozen during winters. During summers, the water-saturated surface soil above the underlying ice of the permafrost went through countless cycles of freezing and thawing. Large rocks were thrust up from the soil. They accumulated in streams or networks of broken rock surrounded by patches of mud that supported only thin soil and meager vegetation (see photographs above and on page 43). The result of this is collectively referred to as patterned ground. Solifluction (soil flow) terraces formed on slopes where freezing and thawing eased the soil slowly downhill through thousands of years. All are periglacial features, meaning they are related to glacial activity. ■

The smooth, rolling upland surface that spans westward toward Lava Cliffs looks like most of the high areas of the park did before glaciers carved deep canyons. The smooth surface continues over the top of the Lava Cliffs, indicating that it was formed after the volcanic eruptions that created the cliff rock about 25 million years ago.

Stone rings and polygons have been thrust out of the ground above Lava Cliffs by multiple cycles of freezing and thawing. *Photo by William C. Bradley*

WESTBOUND

- Distance from
  Deer Ridge Junction
  to this stop
  15.4 miles (24.8 km)

- Distance from Rock Cut
  2 miles (3.2 km)

- Next stop:
  Gore Range Overlook
  1.1 miles (1.8 km)

EASTBOUND

- Distance from Grand
  Lake Entrance Station
  to this stop
  22.3 miles (35.9 km)

- Distance from Gore
  Range Overlook
  1.1 miles (1.8 km)

- Next stop: Rock Cut
  2 miles (3.2 km)

Turn to page 38

Lava Cliffs

GEOLOGY
10
STOP

This stop is a large parking area on the north side of the road. It's on the right for the westbound traveler, on the left for the eastbound traveler (Arrowhead Sign 7).

The cliffs just west of this parking area are composed of layers of volcanic rocks deposited in a depression on the upland, rolling land surface. These rocks are primarily ash-flow rhyolite tuffs that formed when red-hot volcanic ash flowed rapidly downslope from an erupting volcano. The ash contained some crystals of feldspar and quartz that had already solidified in the magma chamber under the volcano, but most of the material was still a viscous liquid. When the semi-molten material stopped flowing, it chilled so rapidly that it solidified into a volcanic glass called obsidian. The resulting rock was

The Lava Cliffs are composed of volcanic rocks deposited during volcanic eruptions in the Never Summer Mountains between 24 and 29 million years ago. These cliffs form the headwall of the cirque of a small glacier that was a tributary of the Fall River glacier. Iceberg Lake occupies the center of the cirque. Small snowbank "icebergs" fall into the lake's water almost all summer.

Rhyolite tuff is formed when red-hot, semi-molten volcanic ash erupts and flows rapidly down the slopes of a volcano. After the flow halts, the heat and accumulated weight cause the viscous liquid to fuse into obsidian glass. This photograph was taken at roadside west of the parking area.

A close-up of rhyolite tuff's polished surface reveals feldspar and quartz crystals that already had formed in the volcanic magma before eruption. These crystals are now surrounded by glassy obsidian.

made up of well-formed crystals surrounded by black or brown glass (see photographs above).

The volcanic eruptions resulted from the collision of Earth's tectonic plates beginning about 130 million years ago. The collision not only lifted the Rocky Mountains from near sea level to their present elevation more than two miles higher, but it also formed magma at great depths.

Stresses from the uplift of mountains pulled the Earth's crust apart along deep faults, allowing magma to work its way to the surface and erupt. These volcanic eruptions broke through the surface on the west side of today's Rocky Mountain National Park about 29 million years ago and lasted about 5 million years. The volcanoes occupied most of the present area of the Never Summer Mountains. Based on their composition, the rocks in the Lava Cliffs probably came from the area of Red Mountain, only a few miles west of here (see the discussion and photograph at Farview Curve on page 58).

At the base of the Lava Cliffs is little Iceberg Lake. Large snow cornices that form on the steep cliffs fall into the lake during the spring thaw. The shade present much of the day prevents the snow and ice from melting until late summer. Iceberg Lake is in a cirque formed at the head of a small glacier that was tributary to the main glacier in the Fall River Valley to the north. There are several areas of hummocky steps around Iceberg Lake, landslides that slid down the steep slopes of the cirque. ∎

Gore Range Overlook

GEOLOGY
11
STOP

This stop is at a large parking area on the south side of the road near a broad curve. The parking area is on the left for westbound travelers, on the right for eastbound travelers.

WESTBOUND

- Distance from Deer Ridge Junction to this stop
  16.5 miles (26.5 km)

- Distance from Lava Cliffs
  1.1 miles (1.8 km)

- Next stop: Alpine Visitor Center
  0.9 mile (1.4 km)

EASTBOUND

- Distance from Grand Lake Entrance Station to this stop
  21.2 miles (34.1 km)

- Distance from Alpine Visitor Center
  0.9 mile (1.4 km)

- Next stop: Lava Cliffs
  1.1 miles (1.8 km)
  Turn to page 44

The elevation at the Gore Range Overlook is 12,010 feet. The view to the southeast sweeps across Forest Canyon to the mountain ridge beyond. The top of the ridge is the Continental Divide, which is discussed in more detail at Milner Pass (see page 55 and the fold-out map). At the far left end of the mountain range is the characteristic flat summit of Longs Peak.

Forest Canyon is a straight valley extending southeastward from Forest Canyon Pass to Moraine Park, a distance of approximately 8.75 miles (14.1 km) (see fold-out map). This valley follows a fault, which weakened the Proterozoic rocks so they could be more easily eroded. The original stream course followed the fault line, creating a straight, V-shaped valley that was later widened and deepened by a series of glaciers during the Ice Age. This is the

Located east of the Continental Divide, Longs Peak is to the far left of the peaks along the Divide in the foreground. Forest Canyon is the straight valley on this side of the mountains. The canyon was occupied by one of the major park glaciers about 20,000 years ago.

upper end of the deep canyon viewed from Forest Canyon Overlook. The Big Thompson River now flows in this valley.

On a clear day, you can see the Gore Range on the far southern horizon at a distance of about 60 miles (96.5 km). The peaks, snowcapped in early summer, reach elevations of more than 13,000 feet (3,962 m). This mountain range near Vail Pass is composed of granite and metamorphic rocks similar to those found in the park.

The view to the southwest spans from low Forest Canyon Pass toward the Never Summer Mountains. These mountains, which in large part are the roots of extinct volcanoes, are dis-

cussed in more detail at Farview Curve (page 58).

To the right, and more westerly, is Specimen Mountain (see fold-out map and the photograph below). Detailed geologic studies have revealed that Specimen Mountain, once thought to be a volcano, is a pile of volcanic rocks that accumulated during the same series of eruptions that deposited the rocks at Lava Cliffs. Specimen Mountain was named for mineral specimens found in cavities and layers in the volcanic rocks, including opal, agate, chalcedony, and black obsidian.

Treeline is easily distinguished on the hillside to the left below Specimen Mountain (see photograph below).

NEVER SUMMER MOUNTAINS

SPECIMEN MOUNTAIN

TREELINE

CACHE LA POUDRE VALLEY

The Never Summer Mountains were the site of volcanoes that erupted between 24 and 29 million years ago. Round-topped Specimen Mountain, once thought to be a volcano, is a pile of volcanic rocks that accumulated from eruptions farther west.

Like chocolate syrup flowing down the sides of an ice cream sundae, the volcanic rocks from the top of the hill west of the Lava Cliffs are flowing inch-by-inch downslope because of the actions of freezing, thawing, and gravity.

Treeline is discussed in more detail at the Alpine Visitor Center (page 50).

The black rocks on top of the hill east of this parking area are broken blocks of the same ash-flow tuff that comprise the deposit at Lava Cliffs.

Streams and lobes of these dark rocks streak the hillside, where the action of freezing and thawing has caused these rocks to move slowly downhill (see photograph above). ■

Mule deer

WESTBOUND

- Distance from
  Deer Ridge Junction
  to this stop
  17.4 miles (28 km)

- Distance from
  Gore Range Overlook
  0.9 mile (1.4 km)

- Next stop:
  Medicine Bow Curve
  0.4 mile (0.6 km)

EASTBOUND

- Distance from
  Grand Lake Entrance
  Station to this stop
  20.3 miles (32.7 km)

- Distance from
  Medicine Bow Curve
  0.4 mile (0.6 km)

- Next stop:
  Gore Range Overlook
  0.9 mile (1.4 km)

Turn to page 47

Alpine Visitor Center

GEOLOGY
12
STOP

This stop is at Fall River Pass. A large parking area is on the right for westbound travelers, on the left for eastbound travelers (Arrowhead Sign 8).

The Alpine Visitor Center and store are perched at 11,796 feet on the rim of the cirque at the head of the Fall River Valley. This cirque was the basin where snow accumulated to form the Fall River glacier (see photograph below). Glacial ice filled this valley as recently as 15,000 years ago. The glacier flowed down the valley to the far end of Horseshoe Park near today's Fall River Entrance Station (see fold-out map). Even though the upper end of this valley has some curves, its cross section is broadly U-shaped (see photograph on page 51).

Most of the sharply cut cirques in Rocky Mountain National Park are on the east or northeast sides of the mountain ridges. Prevailing winds from the west and southwest deposited massive

Snow cornices are the only reminders of the glacier that filled the Fall River Valley (left) as recently as 15,000 years ago. The Alpine Visitor Center, perched on the rim of the cirque, offers exhibits explaining the geology and wildlife of this area.

Once a small stream valley, Fall River Canyon has been deepened and widened by glaciers and water over the last 1.6 million years. Terraces formed by downslope soil movement look like a giant flight of stairs on the hillside near the bottom of the cirque.

amounts of snow on the downwind sides of the ridges. There, the snow accumulated to form glaciers. The resulting landscape of mountain ridges, with their distinctive semicircular scooped-out hollows, has been called biscuit-board topography (see the display in the visitor center). The cirques on the west and southwest sides of the ridges, though not as well defined, produced some of the largest glaciers in the park during periods of maximum ice accumulation.

A stairway on the north side of the parking area leads to the top of the hill.

There, the view is magnificent. If you make the climb up the Alpine Ridge Trail, it will become obvious why this rise has been nicknamed "Huffer Hill." The elevation on the summit is 12,005 feet (3,659 m) above sea level. Exhibits on top show the role of glaciers in shaping this part of the park.

Notice the well-defined upper limit of trees in the valleys surrounding the Alpine Visitor Center. The treeline, also called timberline, ranges from 11,000 to 11,500 feet (3,353 to 3,505 m) elevation in Rocky Mountain National Park. In the Fall River Valley, treeline

Blue columbine | **AZ**

lies at about 11,200 feet (3,414 m) (see photograph on page 51).

The treeline is the upper limit for the growth of trees. This limit is controlled by year-round average temperatures and moisture which, in turn, regulate the length and conditions of the growing season. Immediately below the treeline is a zone of small plants, shrubs, and gnarled trees called *krummholz*, German for twisted wood. The *krummholz* reaches upward into alpine tundra that supports only very small plants. The floor of the cirque below the Alpine Visitor Center is in this transition zone of shrubs separating forest and tundra.

In general, the elevation of treeline is lower the farther one travels away from the Earth's equator. For instance, the treeline on Mount Kenya near the equator is about 14,300 feet (4,360 m). At Glacier National Park near the United States–Canada border, treeline is near 7,000 feet (2,100 m). In the mountains near Anchorage, Alaska, treeline is around 2,000 feet (600 m) above sea level. Few trees grow north of the Arctic Circle, where treeline is very close to sea level. An exhibit in the visitor center illustrates the relationship of treeline to latitude.

Across the cirque, near the skyline, you can see dark layers of volcanic rocks deposited by eruptions in the Never Summer Mountains west of here. These are the same volcanic rocks seen at Gore Range Overlook and in the Lava Cliffs just west of Iceberg Lake (see fold-out map). ∎

Medicine Bow Curve

This stop is in a large parking area at a hairpin turn in Trail Ridge Road on the right side for westbound travelers, on the left side for eastbound travelers (Arrowhead Sign 9).

WESTBOUND

- Distance from Deer Ridge Junction to this stop
  17.8 miles (28.6 km)

- Distance from Alpine Visitor Center
  0.4 mile (0.6 km)

- Next stop: Milner Pass
  3.8 miles (6.1 km)

EASTBOUND

- Distance from Grand Lake Entrance Station to this stop
  19.9 miles (32 km)

- Distance from Milner Pass
  3.8 miles (6.1 km)

- Next stop: Alpine Visitor Center
  0.4 mile (0.6 km)

Turn to page 50

The headwaters of the Cache la Poudre River are just out of sight, up the valley that begins below this stop at Poudre Lake. The river flows down the valley to the right, swings past the round, tree-covered hill in the middle foreground, and then flows away to the north. Eventually, it turns east and leaves the mountains near Fort Collins, Colorado.

During the Ice Age, the portions of the Cache la Poudre Valley visible from here were filled by one of the largest glaciers in the park. The valley to the left and all of the valleys and tributaries to the right contained ice to a level of the present treeline. All of the tree-covered hills visible from here were under ice.

The meadows at the bottom of the

NEVER SUMMER MOUNTAINS

SPECIMEN MOUNTAIN

The headwaters of the Cache la Poudre River to the left of Specimen Mountain, are just out of sight at Milner Pass. The upper limit of tree growth—the treeline—is clearly defined on both sides of the valley.

One of the most massive glaciers in the park area filled this valley with ice to the treeline as recently as 15,000 years ago. The meadow in the valley floor is water saturated and too wet to support tree growth.

Cache la Poudre Valley contain some of the park's major wetlands. Springs flowing into the valley from both sides keep the ground so saturated with water that trees can't grow.

The name of the Cache la Poudre River came from French-speaking fur traders who buried a stash (*cache*) of supplies, including black powder (*poudre*), along the riverbank just west of Fort Collins during an early autumn snowstorm. After recovering their cache the next spring, they continued to refer to this river as the hiding place of the gunpowder—"La Cache la Poudre."

The Medicine Bow Mountains, which extend into Wyoming, can be seen from here at a distance of 20 miles (32.2 km). ■

Milner Pass

GEOLOGY
14
STOP

This stop is in the parking area at the south end of Poudre Lake on the east side of the road. It's to the left for westbound travelers, to the right for eastbound travelers (Arrowhead Sign 10).

WESTBOUND

■ Distance from Deer Ridge Junction to this stop
21.6 miles (34.7 km)

■ Distance from Medicine Bow Curve
3.8 miles (6.1 km)

■ Next stop: Farview Curve
2.2 miles (3.5 km)

EASTBOUND

■ Distance from Grand Lake Entrance Station to this stop
16.1 miles (25.9 km)

■ Distance from Farview Curve
2.2 miles (3.5 km)

■ Next stop: Medicine Bow Curve
3.8 miles (6.1 km)

Turn to page 53

Milner Pass is on the Continental Divide, the great dividing line between areas where rainwater or snowmelt flows toward the Pacific Ocean or the Atlantic Ocean. The Continental Divide, which traverses the tops of high mountain ridges or passes, appears to have dipped into a valley at Milner Pass.

The elevation here is 10,759 feet (3,279 m) above sea level. The apparent low elevation of Milner Pass is deceptive because westbound travelers have been driving for several miles at considerably higher elevations. The fold-out map shows that the Continental Divide is on the top of the mountain ridges south of Forest Canyon and descends to this point to cross the valley. From here, the Divide goes over the summit of Specimen Mountain, across

On the Continental Divide, Poudre Lake forms the headwaters of the Cache la Poudre River. Water flows from this lake toward the Gulf of Mexico.

55

The spires across the lake are pegmatite dikes that intruded metamorphic rocks hundreds of millions of years ago.

This trail starting at the south end of Poudre Lake offers visitors a chance to closely inspect the pegmatite dikes.

Sheep Rock, just down the valley from Milner Pass, was rounded by glacial ice forced to flow up the valley by a massive accumulation of ice along what is now the Colorado River.

La Poudre Pass, and along the crest of the Never Summer Mountains.

To the southwest, down the valley, is a hill called Sheep Rock, which was rounded by glacial ice. Glacial grooves and scratches on the top of this hill and the ice-plucked cliff face on the upstream side indicate that ice from a huge glacier in the Colorado River Valley actually flowed **up** this valley from the west. The glacier flowed over Milner Pass and continued northeastward down the Cache la Poudre Valley (see fold-out map). This glacier is explained in more detail at Farview Curve (page 58).

The two light-colored spires across the lake are dikes of pegmatite injected parallel to layers of the metamorphic rocks about 1.7 billion years ago. The softer biotite schist has weathered away from the harder pegmatite dikes, leaving them standing as monuments. A trail from the north end of the parking area will take you by these spires. ■

WESTBOUND

- Distance from
  Deer Ridge Junction
  to this stop
      23.8 miles (38.3 km)

- Distance from
  Milner Pass
      2.2 miles (3.5 km)

- Next stop:
  Rock Outcrop
      3.3 miles (5.3 km)

EASTBOUND

- Distance from Grand
  Lake Entrance Station
  to this stop
      13.9 miles (22.4 km)

- Distance from
  Rock Outcrop
      3.3 miles (5.3 km)

- Next stop: Milner Pass
      2.2 miles (3.5 km)

Turn to page 55

Farview Curve

This stop is on the west side of a large curve in the road. It is to the left for westbound travelers, to the right for eastbound travelers (Arrowhead Sign 11).

Across the valley in front of you are the Never Summer Mountains. Some of the mountain peaks bear the names of clouds—Cirrus, Cumulus, Nimbus, and Stratus.

The Never Summer Mountains, from Mount Nimbus northward, are composed of granitic rocks that are the remnants of the magma chambers below volcanoes that erupted between 24 and 29 million years ago in upper Oligocene Time (see fold-out map and the geologic time scale on page 4). Beginning about 130 million years ago, collisions of tectonic plates of the Earth's crust produced major uplifts and complex faulting that allowed molten lavas from deep within the crust to move upward to vent at the surface. The volcanic rocks of Specimen Mountain and

The Never Summer Mountains receive abundant winter snow that lasts long into warmer seasons. The mountains to the right of Mount Nimbus are remnants of volcanoes that erupted 24 to 29 million years ago. Red Mountain is the source of volcanic rocks found at Specimen Mountain and Lava Cliffs.

The Colorado River snakes down a valley widened and straightened by one of the largest glaciers in the park.

the Lava Cliffs east of here came from Red Mountain, the cone-shaped peak straight across the valley (see photograph on page 58). Several thousand feet of volcanic rocks have been eroded to produce the mountains you see today.

Numerous mining claims were staked in the late 1800s in the Never Summer Mountains and elsewhere in the park by optimistic prospectors who hoped to make their fortunes digging gold and silver. One of these sites was at Lulu City, about 3 miles (4.8 km) up the Colorado River from here. Although rich deposits of gold, silver, tungsten, molybdenum, uranium, lead, and zinc have been mined in the so-called Colorado Mineral Belt to the south, no minerals of commercial value were found in the area of Rocky Moun-

tain National Park. Although this disappointed prospectors and miners, it helped make possible the setting aside of this area as a national park.

The age difference between the igneous Never Summer Mountains and the Proterozoic metamorphic and igneous rocks farther east in the park is striking. The oldest Proterozoic rocks were formed about 1,700 million years ago; the youngest volcanic rocks were erupted about 24 million years ago—a difference of 1,676 million years. From a geological point of view, these volcanic eruptions were relatively recent events.

The Colorado River flows to the left in the valley below (see photograph above). The headwaters of the Colorado are about 6 miles (9.6 km) up the valley to the right, near La Poudre Pass.

A jumble of huge landslide blocks litters the slopes of Jackstraw Mountain across the valley east of Farview Curve.

The river meanders back and forth across the valley, much as the Fall River wanders through the east side of the park. The meanders result from the sidewise erosion and deposition of stream sediments as the river flows over the relatively flat valley bottom.

When streams flow down a steep valley, the energy of moving water is directed to downward cutting of the streambed. The energetic water can carry eroded material to lower elevations. In a valley that is flat, however, a stream is unable to cut downward, so the energy of the flowing water erodes sideways, first in one direction, then another. As a result, the streams form meander loops that migrate, in time, across the entire valley. These meanderings become more and more convoluted until the channels run into one

another. If a loop is cut off, the stream will leave behind isolated parts that can look like horseshoes. These abandoned loops are called oxbow, or horseshoe, lakes.

The largest glacier in the park was 20 miles (32.2 km) long. It flowed down the Colorado River Valley from the present headwaters of the river. Its terminus was near the south end of Shadow Mountain Reservoir. This glacier received inflow from so many valleys in the Never Summer Mountains, as well as from the head of its own valley, that it became an ice field nearly 2,200 feet (670 m) thick near this overlook.

Part of the glacier was forced to flow **up** the tributary valley to Milner Pass, over the pass, and into the drainage of the present Cache la Poudre river. To illustrate the thickness of this glacier, the upper surface of the ice was as high as the present treeline, which is about 1,100 feet (335 m) above this parking area. The Empire State Building in New York City is 1,250 feet (381 m) high.

The Grand Ditch makes a conspicuous, nearly horizontal, scar across the side of the Never Summer Mountains. This ditch was constructed between 1890 and 1932 to collect and transport water from the Never Summer Mountains across La Poudre Pass, into Long Draw Reservoir, and, finally, into the Cache la Poudre River (see fold-out map). The 35,000 acre-feet of water annually collected by this ditch—enough water to serve the needs of 136,000 households a year—still is used for agriculture on the eastern side of the Rocky Mountains near Fort Collins, Colorado.

A major landslide area on the flanks of Jackstraw Mountain can be seen across the tributary valley to the east of here. It is best seen by walking up the road 200 or 300 yards (see photograph on page 60). The jumbled blocks in this landslide started moving after the tributary glacier in this valley melted between 15,000 and 12,000 years ago. ■

Yellow-bellied marmot

WESTBOUND

- Distance from
  Deer Ridge Junction
  to this stop
        27.1 miles (43.6 km)

- Distance from
  Farview Curve
        3.3 miles (5.3 km)

- Next stop: Bowen-
  Baker Trailhead
        4.3 miles (6.9 km)

EASTBOUND

- Distance from
  Grand Lake Entrance
  to this stop
        10.6 miles (17 km)

- Distance from Bowen-
  Baker Trailhead
        4.3 miles (6.9 km)

- Next stop:
  Farview Curve
        3.3 miles (5.3 km)

Turn to page 58

Rock Outcrop

GEOLOGY
16
STOP

This stop is at a parking area on the right for westbound travelers, just down the road from the last hairpin turn. For eastbound travelers, the parking area is on the left nearly a mile past the turnoff to the Colorado River Trailhead.

Look at the rocks in the outcrop directly across the road from the parking area (see photograph below) and those in the next outcrop up the road. They are metamorphic rocks and pegmatites that rank among the oldest in Rocky Mountain National Park. They're similar to the rocks at Rock Cut near the highest part of Trail Ridge Road. These metamorphic rocks origi-nally were sedimentary rocks inter-spersed with occasional layers of volca-nic rocks deposited in an ancient sea between 1.8 and 2 billion years ago.

The sedimentary rocks were com-posed of layers of shale, siltstone, and sandstone. The volcanic rocks were lay-ers of ash and lavas. When tectonic plates of the Earth collided about 1.7 billion years ago, the sedimentary rocks

Metamorphic biotite schist and biotite gneiss are intruded by lenses and veins of pegmatite. These rocks are among the oldest in the park.

were caught up in an episode of mountain building and subjected to enormous heat and pressure. The shale and siltstone were recrystallized into schist, and the siltstone and sandstone were recrystallized into gneiss (pronounced NICE). Most of these rocks are biotite schist, the thinly layered rocks that contain alternating layers of light and dark minerals.

The dark layers are biotite mica, which contains relatively large quantities of iron. The lighter layers are quartz and feldspar, which in large part were major constituents of the original sedimentary rocks. The biotite schist has undergone several episodes of folding during different stages of mountain building over periods of hundreds of millions of years.

Heat and pressure during metamorphism caused parts of the rocks to melt. The liquid was injected along layers of the unmelted parts of the rocks. The resulting layers and lenses of igneous rock are called pegmatite. Pegmatite is discussed in more detail at Rock Cut (page 38). ■

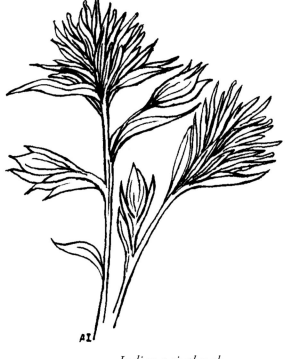

Indian paintbrush

WESTBOUND

- Distance from
  Deer Ridge Junction
  to this stop
      31.4 miles (50.5 km)

- Distance from
  Rock Outcrop
      4.3 miles (6.9 km)

- Distance to the Grand
  Lake Entrance Station
      6.3 miles (10.1 km)

*This completes your geologic tour of
Rocky Mountain National Park*

EASTBOUND

- Distance from Grand
  Lake Entrance Station
  to this stop
      6.3 miles (10.1 km)

- Next stop:
  Rock Outcrop
      4.3 miles (6.9 km) east

Turn to page 62

Bowen-Baker Trailhead

GEOLOGY
17
STOP

This stop is at a parking area for the Bowen-Baker Trailhead, found on the right side by westbound travelers, on the left by eastbound travelers. Park here and walk across the bridge over the Colorado River and westward on the gravel road to the middle of the valley.

The north-south valley of the Colorado River occupies a major fault zone that separates the Never Summer Mountains on the west from the main portion of Rocky Mountain National Park on the east. The rocks of the Never Summer Mountains west of the fault are geologically young. The rocks east of the fault are very old.

The Never Summer Mountains are erosional remnants of volcanic mountains that were several thousand feet higher when they were erupting during the Upper Oligocene Epoch

U-shaped Baker Gulch was carved by tributaries of the huge glaciers that flowed down the Colorado River Valley during the last 1.6 million years. The Grand Ditch flows in the cut high on the gulch's right side.

The rhyolite lava flow on the hillside east of Trail Ridge Road is the result of a volcano that erupted in the Never Summer Mountains about 26 million years ago.

between 29 and 24 million years ago. Most of the present-day Never Summer Mountains are composed of granites that are remnants of the magma chambers under the volcanoes. These volcanoes erupted as a result of uplifting and complex faulting related to the formation of the Rocky Mountains (see geologic time scale on page 4).

Most of the rocks on the east side of the major fault zone are Proterozoic rocks that range in age from 1.4 to 1.7 billion years. They are composed of metamorphic biotite schist and gneiss, and of Silver Plume granite. The metamorphic rocks can be examined closely at the Rock Outcrop (page 62) and at the Rock Cut (page 38). The granite can be examined at Rainbow Curve (page 29).

The valley of the Colorado River was straightened and deepened by many major episodes of glaciation during the Ice Age. The sides of the valley are steep, indicating a general U-shaped profile. The valley bottom now is flat because of sediments deposited in lakes impounded by the terminal moraines of the glaciers. Still more sediments were laid down by the river's many tributary streams after the ice left.

To the west is Baker Gulch, a glaciated, U-shaped tributary valley of the Colorado River (see photograph on page 64). High on the north side of Baker Gulch is the Grand Ditch, a major irrigation project begun in 1890 and completed in 1932 to collect water from the streams along the east side of the Never Summer Mountains.

Though it is a small stream here, the Colorado becomes a mighty river during its 1,400-mile (2,253 km) journey to the Gulf of California. This is the same river that is still carving the Grand Canyon in northern Arizona.

High above the river on the east side of the valley, on the other side of Trail Ridge Road, is an outcrop of rhyolite that flowed from one of the volcanoes in the Never Summer Mountains about 26 million years ago (see photograph on page 65). If you can't see this from where you are standing, you need to walk a little farther across the valley. These volcanic rocks have vertical cracks called columnar joints, which formed when the lava shrank as it cooled.

The two major cooling surfaces were the cool ground surface on which the lava flowed and the upper surface that came in contact with the air. The shrinkage began on both cooling surfaces and progressed to the middle. This is why the cracks are at right angles to the cooling surfaces. The position of this lava flow high above the valley floor indicates that this valley was filled to that level with volcanic ash and lava at the time of the eruption. Erosion by water and ice has since removed most of the volcanic rocks and cut the valley to its present level.

This is the end of the geologic tour for westbound travelers. ∎

Glossary

alluvial fan

A low, relatively flat to gently sloping mass of loose rock material shaped like an open fan, or a segment of a cone, deposited by a stream where it spills from a narrow mountain valley onto a plain or broad valley.

augite

A greenish black to black mineral that contains calcium, sodium, magnesium, iron, aluminum, silicon, and oxygen.

basalt

A dark to medium dark extrusive igneous rock composed chiefly of calcium-rich feldspar and other minerals rich in iron and magnesium.

biotite

A black or dark brown mineral of the mica group composed of potassium, magnesium, iron, aluminum, silicon, oxygen, and hydrogen.

cirque (pronounced SERK)

A steep-walled, amphitheater-like hollow on a mountainside commonly located at the head of a glacial valley. Cirques are produced by the erosive activity of mountain glaciers.

columnar joints

Parallel, prismatic columns, either hexagonal or pentagonal in cross section, in basaltic-type rocks. They form as the result of contraction during cooling.

cornice

An overhanging ledge or a mass of snow or ice on the edge of a steep ridge or cliff face.

crustal plate

See tectonic plate.

daughter products

A series of chemical elements produced by the decay of radioactive elements such as uranium or thorium.

dike

A wall-like mass of igneous rock that cuts through the surrounding rocks. A common example is a pegmatite dike.

dolomite

A mineral ($CaMg[CO_3]_2$) and a rock composed primarily of calcium, magnesium, carbon, and oxygen.

fault

A fracture in rock along which there has been movement ranging from a few centimeters to many kilometers.

feldspar

The most common rock-forming mineral group in the Earth's crust. It is composed primarily of the elements sodium, calcium, potassium, aluminum, silicon, and oxygen.

gabbro

A dark, coarse-grained igneous rock that contains primarily calcium-rich feldspar and other silicate minerals that are rich in iron and magnesium. It is the coarse-grained equivalent of basalt.

glacier

A large mass of ice formed by the compaction and recrystallization of snow. Glaciers move slowly by internal flow and slippage downslope due to gravity.

gneiss (pronounced NICE)

A coarsely foliated metamorphic rock that commonly contains layers of minerals like those found in granite, such as quartz, feldspar, and mica.

granite

A crystalline igneous rock composed chiefly of potassium- and sodium-rich feldspar, quartz, and mica.

hanging valley

A glacial valley whose mouth is at a relatively high level on the steep side of a larger glacial valley. The larger valley was eroded by a major glacier and the smaller one by a tributary glacier.

igneous rock

A rock formed from molten or partly molten rock material. The word is derived from the Latin *ignis*, meaning fire.

isotope

One of two or more types of the same chemical element differing from one another by their different atomic weights.

krummholz

Deformed trees that grow in the transition zone between treeline and tundra. The word is German for twisted wood.

lateral moraine

A ridgelike accumulation of rock debris carried on, or deposited at or near, the side of a valley glacier.

lava

A general term for molten rock at the Earth's surface, as from a volcano. The term also applies to the rock that is solidified from it.

lens

A geologic deposit that is thick in the middle and thinner toward the edges, resembling a convex lens.

limestone

A sedimentary rock composed primarily of the mineral calcite ($CaCO_3$).

limonite

A general term for a group of brown, fine-grained minerals composed of hydrous ferric oxides. This material is the major component in common rust.

magma

Molten rock material resulting from the melting of preexisting rocks deep within the crust of the Earth.

magnetite

A black, shiny, strongly magnetic mineral that contains iron, oxygen, and sometimes magnesium.

marble

Recrystallized limestone or dolomite composed mostly of the minerals calcite or dolomite.

meander

One of a series of somewhat regular, sharp, freely developing and sinuous curves, bends, loops, turns, or windings in the course of a stream.

metamorphic rock

A rock derived from preexisting rocks by mineralogical, chemical, and structural changes in response to changes in temperature and pressure within the Earth's crust.

mica

A group of layered minerals that split into thin, flexible flakes. Mica is composed primarily of the elements potassium, sodium, calcium, magnesium, aluminum, silicon, oxygen, and hydrogen.

microcline

A clear, white to light gray, pale yellow, brick red, or green mineral of the alkali feldspar group ($KAlSi_3O_8$).

mineral

An inorganic substance occurring naturally in the Earth and having a consistent and distinctive set of physical properties and a composition that can be expressed by a chemical formula. Examples are calcite (calcium carbonate, $CaCO_3$) and quartz (SiO_2).

moraine

A ridge, mound, or other accumulation of unsorted and unstratified rock debris carried or deposited by a glacier.

outcrop

That part of a geologic formation or structure that appears at the surface of the Earth.

pater noster lakes

A string of lakes, usually in a glacial valley, that has the appearance of a string of rosary beads. The name comes from the Latin words for "our father."

pegmatite

An exceptionally coarse-grained igneous rock, with interlocking crystals, usually found in dikes, veins, or lenses. The composition is similar to granite, containing quartz, feldspar, and mica.

periglacial

The processes, conditions, areas, climates, and topographic features influenced by the cold temperatures at the margins of former and existing glaciers and ice sheets.

permafrost

Permanently frozen soil and rock resulting from temperatures remaining below freezing for long periods of time. A thin top layer can thaw during brief warming periods.

plagioclase

A group of feldspar minerals that contains sodium, calcium, aluminum, silicon, and oxygen. It is a common mineral in various types of granite.

quartz

The most common single mineral in the Earth's crust. It is composed of silicon and oxygen (SiO_2).

radioactive decay

A spontaneous process in which an isotope (specific atomic weight) of one element loses particles from its nucleus to form an isotope of a new element. The rate of decay is constant, so the amount of decay indicates elapsed time.

rhyolite

A fine-grained igneous rock that contains major quantities of quartz and feldspar. It is the fine-grained equivalent of a granite.

rock

An aggregation of one or more minerals that occurs naturally in the Earth's crust. An example is granite, composed of the minerals quartz, feldspar, and mica.

sandstone

A sedimentary rock composed primarily of sand-sized grains of quartz.

sanidine

A high-temperature mineral of the alkali feldspar group ($KAlSi_3O_8$).

schist

A thinly-layered metamorphic rock, which in Rocky Mountain National Park commonly contains mostly biotite mica, with some quartz and feldspar.

sedimentary rock

A rock resulting from the consolidation of sediment that was transported by water, wind, or ice.

silicate

A mineral or chemical compound which has major components of silicon and oxygen.

slate

A compact, fine-grained metamorphic rock formed from such rocks as shale and volcanic ash. Slate usually splits along planes independent of the original bedding or laminations.

tectonic plates

Segments of the Earth's crust that "float" on a viscous layer below the crust and move more or less independently. These plates encounter each other by colliding or grinding past one another, much like ice floes in a river.

tourmaline

A silicate mineral group that contains a wide variety of elements. It is commonly black.

tuff

A rock formed by the compaction of volcanic ash. It commonly contains crystals of feldspar, sanidine, and quartz.

viscous

Thick, syrupy, and sticky.

Major rock types in Rocky Mountain National Park

Listed below are the major types of rocks you will see on your geologic tour of Rocky Mountain National Park. Their distribution is shown in the geologic cross section on page 8, and on the fold-out map at the back of the book.

Quaternary deposits

Valley alluvium
Sand and gravel deposited by streams on the valley floors after the glaciers melted.

Glacial till
Unsorted mud, sand, gravel, and boulders carried by the glaciers and deposited in, or distributed across, the valley floors when the ice melted.

Glacial moraine
Ridges of glacial till deposited along the margins of the glaciers. Both the moraines and till that are easily seen in the park were deposited by the most recent glaciers, which reached their maximum extent about 20,000 years ago.

Tertiary volcanic rocks

Ash-flow rhyolite tuff
At Lava Cliffs and Specimen Mountain. These volcanic rock formations are composed of crystals of sanidine, quartz, plagioclase, and minor amounts of biotite in a matrix of glass. Much of this volcanic glass has now been weathered to some extent. These rocks erupted from volcanic vents once present in the Never Summer Mountains and accumulated as red-hot mineral crystals and volcanic ash. As the ash accumulated, it fused to form glass. The volcanic eruptions occurred between 29 and 24 million years ago.

Middle Proterozoic rocks

Iron dike
Dark gray to black, iron-rich gabbro (coarse-grained variety of basalt) with limonite staining on joints. Contains the minerals plagioclase, augite, titanium-rich magnetite, and traces of quartz, microcline, and biotite. These rocks intruded the metamorphic rocks and the Silver Plume granite about 1,320 million years ago.

Silver Plume granite
A medium to coarsely crystalline igneous rock containing quartz, feldspar, mica, and minor amounts of other minerals. The Silver Plume granite was formed about 1,420 million years ago when it intruded into and around the earlier-formed biotite schist and gneiss. Silver Plume is the name of the town in the foothills west of Denver, Colorado, where this rock type was first described.

Middle and early Proterozoic rocks

Pegmatite

An exceptionally coarse-grained igneous rock with interlocking crystals usually found as irregular dikes, lenses, or veins in the metamorphic rocks and granite. Two ages of pegmatites in the area of Trail Ridge Road are related to the formation of the metamorphic rocks and to the intrusion of the granite.

Early Proterozoic rocks

Biotite schist and biotite gneiss

Biotite schist is a layered rock containing predominantly a black, platelike mineral called biotite along with fine-grained quartz, feldspar, and a few other minerals. Biotite gneiss is a banded rock with coarser mineral grains. It contains biotite, with larger amounts of quartz and feldspar. These rocks were formed about 1,700 million years ago when this area was the core of an ancient mountain range that was subjected to great heat and pressure.

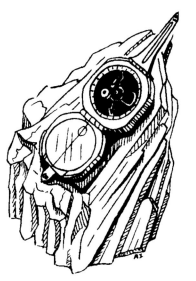

Geologist's compass

References

Arps, L. W., and E. E. Kingery. 1994. *High Country Names*. Boulder, Colo.: Johnson Books.

Braddock, W. A. 1988. *Geologic Cross Sections of Rocky Mountain National Park and Adjacent Terrain*. Estes Park, Colo.: Rocky Mountain Nature Association.

Braddock, W. A., and J. C. Cole. 1990. *Geologic Map of Rocky Mountain National Park and Vicinity, Colorado*. U.S. Geological Survey Map I-1973.

Buchholtz, C. W. 1983. *Rocky Mountain National Park: A History*. Denver, Colo.: Colorado Associated University Press.

Harris, Ann, and Esther Tuttle. 1983. *Geology of National Parks*. Kendall/Hunt Publishing Co.

Osterwald, D. B. 1989. *Rocky Mountain Splendor*. Lakewood, Colo.: Western Guideways, Ltd.

Richmond, G. M. 1974. *Raising the Roof of the Rockies*. Estes Park, Colo.: Rocky Mountain Nature Association.

Willard, B. E., and S. Q. Foster. 1990. *A Roadside Guide to Rocky Mountain National Park*. Boulder, Colo.: Johnson Books.

NEVER SUMMER MOUNTAINS

TRAIL RIDGE ROAD

RO

FC

BBT

	Roaring River alluvial fan			Tertiary intrusives (granites)			Early Proterozoic rocks
	Quaternary alluvium			Cretaceous Pierre shale			Thrust fault
	Pleistocene glacial deposits			Middle Proterozoic granite and pegmatite			Fault line
	Tertiary volcanic rocks (rhyolite and obsidian)			Middle Proterozoic "Iron Dike"			Fault line (conce
						Continental Divi
					BMO	Geology stop

Map labels visible on image:

26, 7, MBC, AVC, GRO, LC, 17, 13, 12, 14, 16, 15, RC, MP, FCO, TRAIL RIDGE ROAD, 11, 22, 23, 24, 25, 18, 19, 20, 21

AVC	Alpine Visitor Center	LC	Lava Cliffs
BBT	Bowen-Baker Trailhead	MBC	Medicine Bow Curve
BME	Beaver Meadows Entrance Station	MP	Milner Pass
BMO	Beaver Meadows Overlook	MPC	Many Parks Curve
DRJ	Deer Ridge Junction	RBC	Rainbow Curve
FC	Farview Curve	RC	Rock Cut
FCO	Forest Canyon Overlook	RO	Rock Outcrop
FRE	Fall River Entrance Station	SL	Sheep Lakes
GRO	Gore Range Overlook		
HPO	Horseshoe Park Overlook		
ID	Iron Dike		

: metamorphic

(aled)

le

1	Moraine Park Museum	3	Deer M
2	McGregor Mountain	4	Morain
		5	North l
		6	South l
		7	Bighor
		8	Roarin
		9	Roarin
		10	Hidden
		11	Tombs
		12	Ypsilo

To Estes Park

To Estes Park

Oblique view fold-out map

Vicinity of Trail Ridge Road
Rocky Mountain National Park, Colorado

Geology simplified from a published map by William A. Braddock, University of Colorado, retired, Boulder, Colorado, and James C. Cole, U.S. Geological Survey, Denver, Colorado.

Digital data produced by the Geographic Information Systems program at Rocky Mountain National Park and the U.S. Geological Survey.

Digital cartography by Carl L. Rick, U.S. Geological Survey, Denver, Colorado.

Drafting by Arthur L. Isom and Dennis L. Welp, U.S. Geological Survey, Denver, Colorado.